"Pittman and Youngs have desc. [barcode obscures text] function of anxiety and obsessive-compulsive diso... quent explanations of solid treatment options and skill building are user-friendly and can be developed with or without a clinician. The language and concepts are down-to-earth and extremely easy to understand. Highly recommended for therapists or anyone struggling with anxiety or OCD."

—**Susan Myers, RN, LCSW, BCD**, employed by the Veteran's Administration Health Care System providing counseling for veterans struggling with anxiety disorders, post-traumatic stress disorder (PTSD), and chronic illness

"Pittman and Youngs have managed to teach complex biological and psychological science in an utterly readable, straightforward, respectful voice that is somehow profoundly calming. The chapter on the language of the amygdala, how it learns and how it communicates, is a perfect bridge between understanding how brain circuitry works in OCD and how to use that knowledge to create the new experiences that can change it."

—**Sally Winston, PsyD**, coauthor of *Overcoming Unwanted Intrusive Thoughts* and *Needing to Know for Sure*, and executive director of the Anxiety and Stress Disorders Institute of Maryland

"If you want to understand the 'what is going on with me?' about OCD and what to do about it, this book is for you. OCD is demystified; the power and control OCD hijacked are given back to the reader. Not only has this book been insightful for me as a clinician, but my clients who read it are more easily able to begin detaching themselves from their intrusive thoughts."

—**Rachel Hiraldo, LPCC, LPC, NCC, CCATP,**
founder and lead counselor at Vivify Counseling and Wellness

"This book is user-friendly and accomplishes the near impossible task of sharing complex neurological concepts using understandable language, helpful descriptions, and practical tips to move past OCD. Patients often begin treatment terrified of their brain, having such uncomfortable and often disturbing thoughts. I am so happy to have a book to help them understand the neuroscience behind OCD, so they can understand OCD from a brain-based perspective. They can shift from shame and blame to self-compassion and empowerment, as well as freedom from the OCD bully within."

—**Debra Kissen, PhD, MHSA**, CEO of Light on Anxiety,
and coauthor of *Rewire Your Anxious Brain for Teens, The Panic Workbook for Teens,* and *Break Free from Intrusive Thoughts*

"*Rewire Your OCD Brain* is the road map you've been waiting for to help you find freedom from your obsessions and compulsive behaviors—an absolute must-read for clinicians and courageous souls alike."

—**Tara Bixby, LPC**, founder of courageously.u

Rewire Your OCD Brain

POWERFUL
NEUROSCIENCE-BASED SKILLS
to BREAK FREE *from* OBSESSIVE
THOUGHTS & FEARS

Catherine M. Pittman, PhD
William H. Youngs, PhD

New Harbinger Publications, Inc.

Publisher's Note

NEW HARBINGER PUBLICATIONS is a
registered trademark of New Harbinger Publications, Inc.

Distributed in Canada by Raincoast Books

Copyright © 2021 by Catherine M. Pittman and William H. Youngs
New Harbinger Publications, Inc.
5674 Shattuck Avenue
Oakland, CA 94609
www.newharbinger.com

Cover design by Amy Shoup; Acquired by Jess O'Brien;
Edited by Rona Bernstein; Illustrations by Arrianna Leigh

All Rights Reserved

Library of Congress Cataloging-in-Publication Data on file

Names: Pittman, Catherine M., author. | Youngs, William H., author.
Title: Rewire your OCD brain / Catherine M. Pittman & William H. Youngs.
Description: Oakland, CA : New Harbinger Publications, Inc., [2021] | Includes
    bibliographical references.
Identifiers: LCCN 2020052569 | ISBN 9781684037186 (trade paperback)
Subjects: LCSH: Obsessive-compulsive disorder.
Classification: LCC RC533 .P497 2021 | DDC 616.85/227--dc23
LC record available at https://lccn.loc.gov/2020052569

Printed in the United States of America

24   23   22

10   9   8   7   6   5   4   3   2

CONTENTS

PART 1

Obsessive Brain Basics

CHAPTER 1

Obsession in the Brain
The Wonderful and Terrible Human Brain

We tend to think very highly of our brains and focus on all that we have achieved with them, from building pyramids to landing on the moon. But our brains can also trap us by creating perceptions, thoughts, and beliefs that torment us. Your brain can produce doubts that worry you continuously. It can dominate your focus with thoughts you can't stop thinking about. It can lead you to feel you must do something over and over in order to feel relief. It can prevent you from making the simplest decision by endlessly producing different scenarios, making it seem impossible to know what's right. How can you break out of these processes that the brain produces?

This may seem to be a very difficult question to answer. After all, you live within the reality produced by your brain. Every sight you see is provided by the intricate connections between your eyes and the parts of your brain that process and interpret the information your eyes receive. If those connections are destroyed, you will not be able to see, even if your eyes are perfectly healthy. Similarly, to hear a sound, you are dependent on your brain to interpret the meaning of sounds that vibrate your eardrum, from the ticking of a clock to the shout of the word "Liberty!" If damage occurs in areas of the temporal lobe, the part that interprets the meaning of words, your friend's words will suddenly sound to you like gibberish or some strange foreign language. Your ears and portions of your brain are processing the sounds so you can hear them, but all the memories of what the sounds mean have been lost because that portion of the brain is damaged. Your brain gives the sounds you hear their meaning.

Most of us do not consider how dependent we are on our brain for our perception of reality, unless we lose it. Our perceptions are shaped by the way the brain works, with numerous parts of the brain contributing to each perception. When Fran was hit by a car and struck her head on the pavement, she damaged a specific area in the back of her brain called the fusiform gyrus. She suddenly found she could not recognize faces. She saw someone's face, but she couldn't tell whether this was the face of someone she knew or a stranger. As soon as the person spoke, she could generally figure out whether it was someone she knew, but for the rest of her life she had difficulties using faces to identify people. She had to rely on other cues like their voices, the topics they discussed, or a specific characteristic like red curly hair or dark eyebrows to help her. Most of us take for granted that our brains are processing the complicated structures of faces and storing detailed memories about them so that we can recognize a familiar face. We don't realize how our reality is shaped by the brain.

Once we recognize that we're living our lives with a brain that processes and interprets our world for us, how do we escape from a brain that is constantly making our world seem dangerous? If thoughts of contamination dominate our thoughts, how do we stop them? If we have worries that we may harm someone or that we have cancer growing in our bodies, how do we escape them? If we feel panic every time we try to drive a car, how do we overcome it? The brain is no longer helping us understand our world and adjust to it. It has turned into a torture chamber!

If the brain has become your torture chamber, how can you escape it? If you have obsessive-compulsive disorder (OCD), you are especially in need of an answer to this question. You need answers about how the brain functions so you can understand how to escape the trap of OCD. The truth is that your brain operates on the basis of specific rules and limitations. Once we understand how the brain works, we can use the rules and processes that govern the brain's functioning in ways that allow us more control over our lives. For example, knowing that the brain can only focus attention on one thing at a time is an important weapon of self-defense for a person with obsessive thinking. We'll show you how to use this limitation in your brain to your advantage, giving you a tool to gain more control over the thoughts and worries that distress you. Once you learn the operations of the OCD brain, as we'll refer to it, you'll understand more ways to have control over it.

Just as you need to understand enough about how your automobile operates to keep it in good condition, you also need to know enough about your brain to get it to operate in a way that keeps you more comfortable. Having obsessions or constant worries in your brain is like having a car with a specific glitch or malfunction. Knowing about the glitch in your car's engine can be helpful in knowing how to manage the problem. You may need to constantly monitor the oil and regularly add some. Both authors of this book have PhDs, but we do not know what to do when a car starts making strange noises or has smoke coming out of the tailpipe. We call Jeremy, our favorite mechanic, and he tells us what needs to be done.

In this book we will explain how you can rewire your OCD brain to improve your life. In the same way we turn to our mechanic, you have come to the right place. We will describe what we know about the processes underlying the obsessive thoughts and compulsions that plague you and give you ways to reduce those thoughts and behaviors and replace them with ones that will allow you to live your life more productively and happily. We will explain the processes that create the torturous feelings of anxiety and dread, feelings that often drive you to perform certain behaviors (sometimes over and over) in order to get some relief. We will teach you how to get out of the vicious cycles you feel trapped in by explaining new ways to cope. This will involve learning ways to deal with obsessive thoughts more effectively as well as finding ways to cope with anxiety—by keeping your brain from producing so much anxiety, living with some anxiety until it subsides, and turning anxiety off as quickly as possible.

What Causes OCD?

If you have OCD, it's normal to wonder what is wrong with your brain. Is there a glitch or malfunction? We want to begin by saying that all of us have glitches and malfunctions in our brains; no one has a perfect brain. Each human being is different and unique, developed from an intricate process of cell union and division that is truly miraculous, combining genetics from two different people. Unlike cars that are tested, designed, and redesigned to minimize problems, this process is not designed to create perfectly operating brains. Every person has certain strengths and weaknesses in his or her brain. The best we can do is to take advantage of the strengths in the brains we have, train our brains to learn as well as we can, and avoid having to deal with aspects of life that require the use of functions in our brain that are less adequate. Some brains are great with mathematics, for example, and will make a person a great accountant, and some brains better stay away from that profession. Some people with OCD have brains that will focus intensely on minute details of a situation for hours, unsatisfied until everything is correct. That kind of brain can make a person a careful and correct writer, but it can also create difficulties in other situations—like never being on time because you don't feel ready to leave the house until everything is correct. We are capable of modifying our brains and behaviors through learning and skill development, but we all have our strengths and weaknesses, many of which are related to the specific brain we inherited.

Parts of the brain that are related to OCD symptoms include areas of the frontal lobes in the cortex, the basal ganglia, and connections between the frontal lobes and the amygdala (Fullana et al. 2017; Nazeer et al. 2020; Welter et al. 2011). We'll be discussing these brain regions in future chapters, but the question here is why are these brain structures reacting differently in those with OCD? We have discovered that OCD and other anxiety-related disorders seem to run in families, suggesting that they are inherited in some way. If you have an anxious parent, you are more likely to have anxiety difficulties. Adoption studies have clarified that it is not just about being raised by an anxious parent—a genetic component definitely contributes to these disorders (Gregory and Eley 2007). But genetics is not the only important factor. We also learn our specific fears and worries from our individual experiences, families, and friends. In other words, the OCD-related processes in your brain are a result of both genetic tendencies and life experiences (Nestadt, Grados, and Samuels 2010).

In addition, researchers have discovered that obsessive thinking and other anxiety symptoms can be acquired through viruses that impact the brain. Two viruses, which typically impact children between ages three and twelve, are called PANDAS (Pediatric Autoimmune Neuropsychiatric Disorder Associated with Streptococcus) and PANS (Pediatric Acute-Onset Neuropsychiatric Syndrome). These viruses cause damage in the basal ganglia, leading to symptoms like obsessive thinking, compulsive behavior, and tics (Bernstein et al. 2010). When a child develops OCD from PANDAS and PANS, the symptoms are likely to come on quickly after the child has the virus, without prior indication that OCD is a problem (Baj et al. 2020).

Clearly, structures in the brain are creating OCD symptoms, whether those structures have been affected by genetics, illness, or learning experiences. What's most important for you to know is that it is not your fault that OCD developed, and you should not blame yourself for that. On the other hand, it is your responsibility to learn about what you're dealing with and how to do something about it, because you deserve a better life than one that is limited by OCD. Symptoms of OCD can be very disruptive in your life, and understanding how they are created by your brain can be helpful in reducing their impact.

Obsessive Thinking in OCD

When you have thoughts that seem to repeat in endless cycles, you are dealing with *obsessive thinking*. Obsessive thinking comes in various forms. The most common kind is *worry*, which refers to thoughts about what could go wrong and potential negative outcomes that could occur. Worries often begin with "What if…" and end with frightening ideas of what might happen. Worries generally change day to day, depending on the circumstances. People often actively and deliberately engage in worry, although some find worry difficult to control. People with OCD can find that worry tends to become time-consuming and uncontrollable (Clark 2020).

Obsessions are another kind of obsessive thinking. Obsessions also involve repetitive thinking, but they are more stable than worries, which often change day to day. For example, Bruce continually focuses on the specific idea that he might not be able to complete his college degree, even when he is getting good grades. Obsessions are unwanted thoughts that can't be dismissed. Obsessions might be worries, but they can also be images or impulses. For example, the image of her little brother being hit by a car pops into Rose's head repeatedly, and she can't get it to stop. Terrence often has the impulse to drive in front of a semi-truck when he is driving on the freeway, and he worries about what it means, even though he never acts on it. Obsessions can also be doubts you have about whether you did something, or a continual focus on deciding what you *should* do. You may continually worry about whether you left your electric hair straightener on or take more than an hour to decide what kind of tea you should drink. Obsessions can be images of what terrible dangers could befall you or the ones you love. Sometimes they are expectations for how you "should" do things that become rigid commands you feel you must obey. Some obsessive thoughts involve situations that are very unlikely or implausible but still cannot be dismissed, like the idea that you ran over a dog on the way to work but didn't notice doing so.

Worries can be about all kinds of situations and topics. The human brain can probably craft a worry to fit almost any situation. Just think about what could go wrong in that situation and you have a worry. Obsessions are more likely to fit into certain categories or themes. One theme is *contamination*, involving worries about being exposed to dirt, germs, viruses, and so on. For example, Gladys is terrified of eating anything at the church potluck because she doesn't know where and how the food was prepared. Another common theme is *orderliness*: objects, events, and even movements need to be correctly arranged or follow a specific order. The contents of drawers, the cars in the driveway, the holiday dinner, even hand movements when talking can become a focus of concern. *Violence and aggression* can be a theme of some individuals' obsessions. A new recruit on a college football team may have a continuous worry that he has harmed someone during his high school career, even though he does not recall doing so. Sometimes people experience impulses or images of performing some aggressive or sexual act and worry that they may do so, even though the very thought distresses them.

Perfectionism is a very common theme of obsessi‚
with OCD. Individuals with perfectionism are focuset
in their appearance or mistakes in their behavior, despi‚
human beings, they cannot achieve perfection. One youn‚
tolerate her piano lessons because, after the instructor de‚
assigned musical piece, the girl could never hope to play i‚   ‚er
lesson the way the instructor did. A graduate student in English was paralyzed in writing her dissertation because no sentence that she wrote in her
first draft seemed to measure up to her idea of "dissertation level" writing.

Other people focus on the sensitive themes of *religion* or *sexuality*. A
man may worry that, since God is aware of all his thoughts, he has no
chance of getting into heaven, and he may constantly seek ways of atoning
for the angry thoughts he had about his parents as a child. Or a woman
may worry that she is a lesbian, even though she has no attraction to
women. We don't know why certain people focus on certain themes, but
the same brain process underlies the different obsessions—it's only the
theme that differs.

Obsessive thinking can dominate a person's life. A college student
may worry that his girlfriend does not love him and call her for reassurance multiple times a day. A mother driving to work might imagine her
home in flames, fearing she left the stove on, and turn back to check,
making her late for work every day. A nanny may worry that she'll grab a
knife from the wooden block on the counter to stab the children as they
sleep, leading her to quit her job. The same distressing thoughts keep
returning again and again. They tend to take up more than an hour each
day, sometimes dominating the individual's waking hours. Obsessive
thoughts often drive the person to take action to cope with the thoughts
or neutralize them in some way, and this often leads to compulsions.

Compulsions are repetitive behaviors or mental acts that a person
engages in to respond to a dreaded thought or situation or to reduce distress. They can take a variety of forms, including checking, counting,
cleaning, and seeking reassurance. Compulsions can come to dominate
your life as much as obsessions because they give you a quick fix that helps
you avoid obsessions or worries and the stress they create. When Damien
gets nervous when in a traffic jam on the highway, he starts mentally

9

ng the numbers on the license plates of cars in front of him. He knows it doesn't make sense, and may even be dangerously distracting, but he feels it protects him in some way. Perhaps you feel a twinge of anxiety about the disarray of the pill bottles in the medicine cabinet, but when you organize them, you feel the anxiety go away. When you carry out a compulsion, the feeling of anxiety associated with certain thoughts or situations decreases, and the relief that you feel rewards you for engaging in that behavior. That relief makes the compulsion stronger. So what is the problem? The problem is that your relief is only temporary. You will soon need to repeat the compulsion again and again... We will discuss how to reduce time-consuming compulsions in chapter 9.

Obsessions and Worries in the Brain

The human brain was developed millions of years ago, in a completely different world from the one we live in now. The pathways and processes in our brain developed to allow humans to survive in a world where they were not only hunters and gatherers, but also prey. As humans learned to farm and build communities, the pathways and processes in our brain adapted to be used in new ways, but that process of adaptation was not what you might expect. Sometimes people discuss the brain's adapting to new demands and opportunities as if it were being remodeled like a home, with new rooms and staircases being built and new appliances installed. Nothing could be further from the truth. The pathways and processes in the brain have remained mostly the same across centuries, like an old log cabin that had a wall removed but kept the same furniture and appliances from a century ago.

Over the centuries, some parts of our brain became larger and more complexly organized, but if you compare our brains—not just to the brains of early humans, but even to the brains of other animals—you find many of the same pathways and processes. Although we definitely have pathways and processes that don't exist in other animals, primarily in our frontal lobes, all the new pathways are connected to pathways that have been operating basically unchanged since we lived as hunter-gatherers. A useful analogy might be that the old cabin may now have a lean-to built onto it to use as a library, with many modern books on the shelves, but the cabin itself still has the old fireplace and all the rough-hewn wooden furniture it once had. Despite the knowledge contained in the books shelved in the library, no modern kitchen or heating system has been built. In a similar way, many parts of our brain have not been remodeled to adapt to our current lives. Primarily in our cortex, we have some additional circuitry that provides new capacities, higher-level thinking, and increased ability to store knowledge, but the basic structure and functions in the brain have remained unaltered. This means that even though we have abilities to think in ways that other animals do not, these parts of our brain are connected to older parts of our brain that operate as if we still were potentially prey for dangerous animals. We have an elaborate defensive system woven into the neural pathways of our brain and body that still impacts our daily responses to threat in a variety of ways. This defensive system is involved in creating the symptoms and distress associated with OCD, and, instead of helping us, the more recently evolved thinking abilities in our brain can actually produce thoughts and images that make the brain *more likely* to activate the defensive system.

Worry and obsessive thinking are some of the thinking abilities that have their roots in this newly developed part of our brain. The lobes of the brain behind our forehead, appropriately called the frontal lobes, became larger as humans evolved as a species and gave us new capacities that other animals do not have. One of these capacities is *anticipation*, which is the ability to think about what *may* happen, before it happens. This ability developed in an area of the brain that controls our movements (Leaver et al. 2009), which makes sense. If the frontal lobes are about making movements in our arms and legs, then developing the ability to anticipate what might happen would help us make appropriate movements. Developing the ability to anticipate before moving would be important in hunting and reacting to physical threats.

But anticipation in the human brain went beyond just movement-related anticipation and developed into an elaborate process, ultimately including the human ability to imagine events and objects that one has never seen. This ability, combined with the ability to *plan* (also rooted in the frontal lobes), led to all kinds of uniquely human behavior—from learning to grow our own food to building remarkable structures, ultimately resulting in scores of amazing activities imagined and ultimately planned by humans—including traveling to the moon.

This amazing ability of the human brain to anticipate is a gift to us in many ways, but it also has a cost. We're able to imagine a range of troubling events, from thinking about breaking an important object to imagining the manner of our own deaths, thoughts that represent events that may or may not occur. And these thoughts take a toll on the human race. We can worry and obsess about countless ideas and images, and, as we do, we experience emotional reactions that are similar to what we'd feel if the event actually occurred. This is because the new parts of the human brain that create these thoughts and images are wired into the rest of the brain. Unfortunately, the rest of the brain has not been remodeled in a way that would allow it to react differently to imagined dangers than it would to real dangers. When you imagine an event, defensive systems in the brain often react as if that event were occurring, preparing you to react physically and giving rise to changes in other parts of the brain and body that your conscious brain experiences in a very emotional manner. Understanding how different parts of the brain and body are connected can help you understand what is happening when you are being tormented by worries and obsessive thoughts or driven to engage in repetitive compulsive behaviors. More importantly, this understanding can also show you how to get out of the vicious cycle that OCD can create.

Pathways in the Brain That Create OCD

Obsessive thinking and worries are produced in the *cortex* of the brain (see figure 1), the large, convoluted, gray structure that fills the topmost part of your skull. One of the main roles of the cortex is to process *sensory information* from our sense organs, like our ears and eyes, so that we understand what we hear and recognize the meaning of what we see. The cortex is instrumental in helping us decide whether what we see, hear, smell, and so on, is something that should concern us. And the cortex doesn't just help us recognize what is going on around us—it also helps us observe our own behavior and detect whether we have made a mistake. The cortex is also where bodily reactions are felt and movements are initiated. These processes in the cortex are essential for responding to signs of danger in our lives. But, if danger is detected by the cortex, it's not only the cortex that responds. Other parts of the brain also become involved in mobilizing a response to the danger.

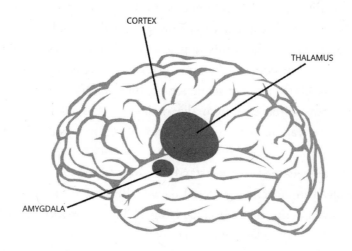

CORTEX

THALAMUS

AMYGDALA

Figure 1

A separate part of your brain, the *amygdala* (pronounced uh-MIG-dull-ah), initiates the body's most effective response to danger. Surprisingly, the amygdala can sometimes recognize dangers *before* the cortex. This means that, in some cases, the amygdala is capable of making your body react to protect you before you even know what you're seeing or hearing. You may even remember some examples of this occurring in your life. One common example is driving on the highway and reacting to a danger so quickly that you have taken an action (turned the wheel rapidly or hit the brake) before you even had time to realize what you were reacting to.

The reason you can react before you even know what is happening is because there are two different pathways in the brain that process infor-mation about potential threats, and they operate on different speeds. Both pathways begin when information from the senses comes through various pathways to the *thalamus*, a walnut-shaped structure in the center of your brain, which determines where to direct the information coming in from the senses. One pathway, which we will call the *amygdala pathway*, processes information about potential threats very quickly, going from sensory pathways to the thalamus and directly to the amygdala. This allows the amygdala, which is close to the thalamus, to have access to sensory information rapidly. The second pathway starts the same as the first, with sensory pathways carrying information to the thalamus, but we call this the *cortex pathway* because in this pathway the thalamus sends the sensory information up into the cortex to process it. Then, the amygdala, which has multiple connections up into the cortex, can have access to this information. The cortex pathway requires more time because the sensory information needs to be sent from the thalamus to the cortex and then processed by various areas of the cortex before it is accessible to the amygdala.

Why do you need to understand the two pathways in the brain that are involved in detecting potential danger? Because these pathways are very involved in the processes underlying OCD! We are fortunate that in the last several decades a great deal of research has explored how the human brain recognizes and responds to potential threats. Research on the neurological underpinnings of fear and anxiety has been conducted in laboratories around the world (Dias et al. 2013). Researchers have identified structures in the brain that detect threats and initiate protective responses in both humans and animals (LeDoux 2015). New technologies, such as functional magnetic resonance imaging (fMRI) and positron emission tomography (PET), have shown us images of how different parts of the human brain respond in a variety of situations. This emerging knowledge allows neuroscientists to assemble a clear picture of the processes that lead to fear and anxiety in the human brain, providing an understanding that surpasses our understanding of all other human emotions. When you have OCD, fear and anxiety are the exact feelings that you need to understand.

In short, the cortex produces thoughts and images, and the amygdala initiates many of the emotional and physical aspects of anxiety and worry that you experience in the process of dealing with OCD. A basic understanding of the two pathways to anxiety will help you understand how and why your brain produces and maintains distressing, intrusive thoughts and learn strategies to get out of the endless cycle of OCD that is limiting your life.

The Self-Defeating Cycle of OCD

Due to the cortex pathway, thoughts in the cortex are capable of activating the amygdala, which creates the reactions we experience as stress and anxiety. It may be helpful to imagine the amygdala "watching" what is going on in the cortex, much like a child might watch television. Just as a child might react to a frightening movie with fear, the amygdala responds to frightening images or thoughts produced in the cortex by creating a defense response in the body, which we experience as fear and anxiety. This, in turn, leads to even greater concern about the thoughts. Basically, the frightening thoughts result in a frightening bodily reaction, and this produces anxiety and increases the focus on the frightening thoughts. You can see how this process can become a cycle, where thoughts lead to anxiety and anxiety creates more concern about and more attention to your thoughts, which leads to even more anxiety.

It's difficult to take your mind off something that seems threatening. In fact, it's the amygdala's job to help you stay alert to potential danger, so the amygdala becomes involved in keeping you focused on these intrusive thoughts. Now you're not only dealing with the intrusive thoughts but are also having to cope with anxiety. You begin to focus on ways of reducing anxiety rather than ways of reducing the intrusive thoughts. Once you understand the role of both the amygdala and the cortex in this cyclical process, you'll use your knowledge of the two pathways to get to the root of the problem.

In summary, OCD is a self-defeating cycle that begins with distressing thoughts that then produce anxiety, which makes the thoughts even more distressing, and turns the problem into a battle against anxiety. You may develop compulsions to help reduce the anxiety, resulting in full-blown

OCD. In this book, we will help you understand the different roles that the cortex and the amygdala play in this self-defeating cycle. Once you know how the different pathways in the brain create distressing thoughts, anxiety, and compulsions, you'll have a much better understanding of what you're fighting, and you'll be ready to learn the strategies that will allow you to win the battle against OCD.

Can I Change My Brain?

In just the last three decades, scientists have learned that the brain has an amazing capacity to change its structures and reorganize its patterns of responding. This is known as *neuroplasticity*. Your brain is not fixed and unchangeable, as so many people, including scientists, assumed in the past. The circuits in your brain are not determined completely by genetics; they are also shaped by your experiences and the way you think and behave. No matter what age you are, you can remodel your brain to respond differently. There are limits, of course, but there's also a surprising amount of flexibility and potential for change in your brain. This includes changing a brain that has a tendency to create intrusive, distressing thoughts that lead to compulsions.

New connections in the brain often develop in surprisingly simple ways you may never have thought of: Aerobic exercise has been shown to promote changes in the brain's anatomy and physiology that are accompanied by measurable changes in mood (Swain et al. 2012). Certain medications help promote growth and changes in circuits of the brain (Drew and Hen 2007). Psychotherapy has also been shown to produce changes (Linden 2006), reducing activation in one area of the brain and increasing it in others. In particular, people who have been taught mindfulness meditation show changes in the connections between areas in the brain that influence anxiety (Yang et al. 2016). In later chapters, we'll explain how these methods can be used to help you in your battle against OCD, but for now, rest assured that, yes, you can change your brain!

We know quite a bit about ways to change the brain to reduce distressing thoughts, to create them less often, and to respond to them differently.

Also, we have learned what techniques are most effective in reducing anxiety for good, not just temporarily. We will teach you to interrupt the OCD brain's patterns of responding and teach the OCD brain to respond according to new patterns by building new neural structures that support more anxiety-resistant responding. This is the process that we refer to as "rewiring" your OCD brain. Of course, the brain has no real "wiring" in it, but surprisingly enough, different neurological structures in our brain do use electrical signals to communicate, and they transmit messages to each other through connections that work very much like wiring. Understanding how the different parts of the brain communicate will allow you to have much more influence over this communication, enabling you to exert more control over these processes than you could before. Also, when you understand how specific parts of the brain learn and form new connections, you will understand how you can rewire, or teach, the different parts of your brain to react in new ways.

In the strategies explained in this book, we will help you use the neuroplasticity of the brain, along with a knowledge of the two pathways that produce anxiety, to make lasting changes in your brain that give you more control over your life. Our goal is to help you not just manage distressing thoughts in a given moment, but also make changes in the way your brain typically responds so that these thoughts become less troublesome in your life. You can also use the information in this book to transform your brain's circuitry so that your brain resists anxiety rather than creating it. Our intention is to give you the knowledge you need to escape from the torture chamber that your OCD brain can create.

CHAPTER 2

The Root of Anxiety
Understanding the Amygdala

We have seen how anxiety in the brain can worsen the process of obsession and lead to a self-defeating cycle. In this chapter, we examine the root of anxiety in the brain, the amygdala. Remembering that the brain is a physical organ, governed by physical processes, will give you an advantage in coping with your OCD difficulties. You will benefit from recognizing that your anxiety is rooted in a physiological process designed to defend you against threats. The processes that give rise to our experience of fear and anxiety begin when the activated amygdala produces a defense response to something identified as a threat. Knowing that the defense process initiated in the amygdala is what leads to your experience of anxiety is key in overcoming the effects of anxiety in OCD. Reducing your anxiety by focusing on its roots in the amygdala will be more effective in the long term than compulsions are, and understanding the amygdala's response will also help you respond differently to your anxiety and obsessions.

Your amygdala, and not your thinking processes, is responsible for creating the physical responses in the body and brain that lead you to experience anxiety. Whether you have a pounding heart or feel nauseous, shaky, or dizzy, you can thank your amygdala. If you understand how and why the amygdala is creating these physical responses, you'll be able to accomplish two important achievements: interpret the meaning of your anxiety more accurately, and have more control over your experience of anxiety. Our goal is to reduce both the *frequency* of your anxiety and your *distress* about it. This will help you get the upper hand when it comes to both obsessions and compulsions.

The amygdala (which is the size and shape of an almond and gets its name from the Greek word for almond) is the root of the defense response we sometimes call the fight-or-flight response. Actually, we have two amygdalae, both located in the center of the brain, but it has become tradition to refer to the amygdala in the singular, and we will follow that tradition here. The differences between the two amygdalae are not essential to our purposes, so we will not be addressing them.

The location of the amygdala in the center of the brain puts it in close proximity to some pretty influential parts of the brain, including the hypothalamus, which gives it the ability to make quick changes in the functioning of your heart, lungs, muscles, and even digestive system. When your amygdala detects a threat of some kind, it is positioned to put into motion a very complex response in your body to allow you to protect yourself. The amygdala can create this defense response before you are even aware of a danger, because it happens before the cortex is even finished processing the sensory information that the amygdala is responding to. Being aware of the cause and purpose of this defense response can help you have a better understanding of what you are experiencing as you try to manage your stress and anxiety.

The Amygdala as an Early Warning Alarm System

Natural selection has made the human amygdala a threat detector whose primary goal is protecting you. (We should note that the amygdala has other roles as well, including in *positive* emotional reactions and aggression, but that's not the focus of this book.) As you go about your day, your amygdala is vigilant for anything that might indicate potential harm and is ready to initiate a defense response to ensure your safety. The way in which the brain is organized allows the amygdala to serve as an early warning system and defense coordinator, which has protected our ancestors from danger for centuries.

Here we will take a close look at the role of the amygdala in the two pathways we described in chapter 1. You'll recall that in the amygdala pathway, sensory information comes in to the thalamus and then is sent directly to the amygdala. In the cortex pathway, incoming sensory information is sent by the thalamus to the cortex before being accessible to the amygdala. These two pathways in the brain each result in the experience of anxiety that fuels both obsessions and compulsions, so it is important to trace both pathways. (See figure 2 for a simplified version of each pathway.)

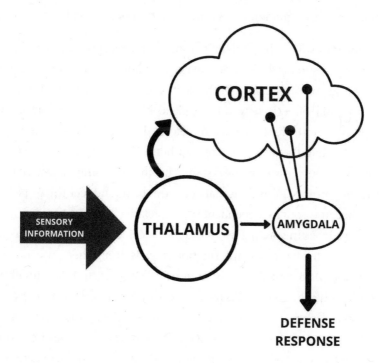

Figure 2

Cortex Pathway: When information is received through the eyes, ears, or other sense organs, it needs to be processed in the cortex pathway in order for you to perceive it. From the sense organs, information travels to the thalamus in the center of the brain. (See figure 1 in chapter 1.) The thalamus acts like an office receptionist, receiving information from the various sense organs and directing the information to wherever it needs to go to be processed in the cortex. For example, the visual information that comes from your eyes as you read this book is sent first to the thalamus, and then the thalamus sends it on to be processed at the back of your head in part of your cortex called the *occipital lobes*. In contrast, sensory information coming from your fingers as you pet a cat in your lap comes up your spinal cord into your thalamus, which then sends it up to a part of your cortex called the *parietal lobes*, where touch information is processed. Only after the information is processed by the occipital lobe or the parietal lobe do you see the words you are reading or feel the soft fur of the cat. Luckily this transmission of information takes only a fraction of a second.

Amygdala Pathway: The thalamus, that neural receptionist, also has a little safety measure. At the same time the thalamus sends sensory information to the correct location in the cortex to be processed, it also sends the information directly to your amygdala. Notice in figure 1 that the amygdala is close to the thalamus; therefore, the amygdala gets the information faster than the other parts of your brain. The amygdala is in a good position to be your early warning system, quickly processing information to assess its importance. Your amygdala can see, hear, and feel things before you can, because you are dependent on the relatively slower processing of the different lobes in your cortex in order to receive information.

In the cortex pathway, information travels from the sense organs (eyes, ears, skin) to your thalamus, then to your cortex, and then finally is processed by the amygdala. This pathway is the slower of the two pathways but still takes less than one second. What *you* see, hear, and feel is based on the cortex, so what you experience from your senses comes from this pathway. In the amygdala pathway, which also can result in the experience of anxiety, information travels from the sense organs to your thalamus and then directly to your amygdala within *milliseconds*. The amygdala is constantly monitoring what you are experiencing more rapidly than you can. Because you depend on the cortex for your information, you don't see or hear exactly what sensory information the amygdala processes. You don't understand why your muscles are tense or why your heart is pounding, but you do experience the bodily reactions, or *defense response*, that the amygdala produces if it detects a potential danger.

Let's look at one of the most common situations in which you may have experienced these pathways in operation. Imagine that you are driving down the freeway, and another car suddenly veers into your lane. Before you have time to think, you have jerked the wheel to the left and driven your car into the grass on the side of the highway. You hit the brake and stop the car and, in your mind, you review what just happened. You feel almost as if *someone else* had jerked the wheel to the left, because you have no recollection of thinking what to do; you just did it. Luckily, that was the right thing to do and you avoided a collision. Or perhaps we should say that *your amygdala* avoided the collision. Consider the two pathways we described above. When the car veered into your lane, the information from your eyes went to the thalamus, which then sent it to the occipital lobe at the back of your cortex. But this information took a bit of time to be processed. At the same time, the amygdala received information about the approaching car directly from the thalamus and processed it more rapidly. Your amygdala was able to detect the danger a fraction of a second before your cortex (or you) detected it. Before you had really even seen the car veer into your lane, your amygdala was able to activate the reaction of jerking the wheel to the left. The swift reaction of your amygdala could have saved your life!

It's worth noting that the reaction your amygdala created was not only your action of swiftly turning the steering wheel. You probably felt a rush of adrenaline and a quick increase in your heart rate at the same time that you were making the quick movement to avoid the danger. People often assume that a reaction like this is simply a reflex...but this was not a reflex, which is based on messages from your spinal cord. A reflex is involved when you burn your fingers, and the signal of the pain gets to your spinal cord, which sends back a signal to pull your hand away from the heat. Reflexes happen *before* the message gets to your brain. In this case, the visual information in your brain needed to be processed by your amygdala for you to jerk the steering wheel; therefore, it was not a reflex. Your amygdala's detection of an oncoming car made your swift defense response possible. You acted before your cortex had time to detect the problem and decide what to do.

Considering the Costs...

You can see the usefulness of the amygdala's ability to see or hear potential dangers more swiftly than the cortex can. The way our brains are organized to give the amygdala quick information has provided the opportunity for rapid reactions that saved our ancestors from countless dangers. But getting information quickly comes with a cost. Through this direct pathway, the amygdala is depending on raw, unprocessed sensory information, not the detailed information you get when the cortex processes the information. Your cortex is able to process and interpret situations more thoroughly than the amygdala, giving it the ability to recognize important details that the amygdala may not. So your amygdala is reacting to a rough idea of what was heard or seen and acting swiftly on the basis of that rather incomplete information. You can think of what the amygdala sees as something like a partially developed, blurry Polaroid picture—it has four legs...but is it a cow or a dog? Luckily, partial information is often enough to detect whether or not there may be a danger. And when swift reacting is not an issue, there is time for the amygdala to get more accurate information from the cortex, as soon as the cortex processes it.

Joseph LeDoux (1996) provided a classic example to illustrate the relative accuracy of these two separate pathways. He proposed that if you are walking in the woods and you suddenly see a brown, snake-like, curved object on the path ahead of you, the amygdala may trigger a defense reaction and you may jump back with your heart beating very fast. But almost immediately, you recognize—based on the processing in your cortex—that the brown object is a piece of vine rather than a snake. You feel a flush of relief, although your heart is still pounding. What happened?

The amygdala detected a potential danger—the raw information from the thalamus indicated the object looked like a snake—and took charge to create an appropriate defense response, except it wasn't appropriate in this case. You jumped back unnecessarily, and also probably experienced an adrenaline rush and increased muscle tension. When the information was processed by the cortex a fraction of a second later, you recognized details that the cortex could provide and saw the object as something that did not pose a danger. Information from the cortex eventually got to the amygdala, so the amygdala could stop raising the alarm. In terms of safety, we are better off as a species if our brains swiftly protect us from potential dangers—and sort out later whether it was necessary or not. Better to be safe than sorry is the amygdala's motto.

An important lesson should not be missed here: *The amygdala's reaction to what you are experiencing may not be correct and may not be needed.* As you're learning more about the amygdala, you realize that it is not making careful, information-based, logical decisions but is actually reacting on the basis of incomplete information, with a bias toward reacting even when it's unnecessary. Nevertheless, because the amygdala has the ability to create a whole variety of bodily reactions on the basis of its interpretation, many people misinterpret those bodily reactions to indicate a true threat. Once you're feeling anxiety as a result of the amygdala creating the defense response, it's easy to *assume* that there's a good reason for the anxiety you're experiencing. This happened when Paulette was carefully choosing her seat in an auditorium one day. As she decided on a specific seat, she sat down and saw what looked like a spider on the arm of her seat. In a fraction of a second, she realized that the shape was simply a knot in the wood that looked somewhat like a spider. (Her amygdala had reacted erroneously, but her cortex could provide more detailed and correct information.) Nevertheless, because Paulette's amygdala had produced the defense response, and her heart was pounding, she couldn't shake the feeling that she shouldn't sit in this chair. *Based on how she was feeling*, she decided to find a new chair in order to be certain she was safe. In other words, she was acting on the basis of the amygdala's erroneous reaction.

In OCD, it's very common to overestimate threats. The amygdala plays a role in this problem. Many times people overestimate threats because they assume the accuracy of the reaction produced by the amygdala. Once you recognize the amygdala's tendency to react in error, you can learn not to always trust your experience of anxiety. Your experience of anxiety is very real, but it's often based on an unnecessary, erroneous defense reaction from the amygdala.

In summary, our brains have a built-in defensive system, which has not been modified much for millions of years and is pretty similar to the neural system in the brains of other animals. When triggered, the amygdala creates a defense response by producing very real and significant changes in our bodies. When the amygdala produces this set of bodily changes, we frequently experience the emotional reactions of fear and anxiety. But these emotional reactions should not be trusted to always indicate danger. Simply feeling fear or anxiety does not validate that a danger exists because the amygdala does not always have an accurate interpretation of the situation.

Defensive Options: The Fight, Flight, or Freeze Response

Let's look in some detail at the swift defense response that the amygdala can create. This response is often called the fight-or-flight response, but research has suggested that it should more appropriately be called the *fight, flight, or freeze (FFF) response* (LeDoux 1996). That's because, when it detects potential danger, the amygdala immediately sends signals to various areas in the brain to help us respond, and the responses the amygdala can select from fall loosely into three categories: fight, flight, or freeze. We can fight, whether that means hitting, pushing, or struggling against someone or something. We can also flee the situation, which includes behaviors like running, ducking to dodge an object, or avoiding situations. Finally, another option is freezing, or being still or immobile. At different times in your life, you may find yourself experiencing each of these types of responses to threatening situations. You may also notice that one or two approaches are more likely for you, noting that you tend to escape and avoid more than you tend to fight, or vice versa. The point is that these general responses are wired into us as preprogrammed defense responses that the amygdala can trigger very quickly and adapt to the given situation.

Consider that, when the amygdala detects a potential threat, its goal is to prepare us either to fight, flee, or freeze. To accomplish this, it must put into action a whole set of physiological responses. When the body needs to move quickly or with a lot of force, the amygdala can swiftly make necessary changes in the body to allow these actions to occur. Fighting or fleeing often requires increased blood pressure and blood supply from the heart. Increased muscle tension and reaction is needed, so glucose is released into the bloodstream to fuel responses in the muscles. At the same time, digestive processes are slowed because blood is rerouted to muscles in the extremities, so you may feel a queasy feeling in your stomach as this occurs. All of these reactions need to occur to prepare the body to react to a threat, and the amygdala can activate all of them. Understanding the actual purpose of these physiological reactions can help you understand them when you experience them. "My heart is pounding because the amygdala is getting me ready to fight." "My legs are shaking because my amygdala is tensing up my muscles to get me ready to run away." The truth is, these reactions often don't help us cope with the stresses that we are facing.

Consider Ruth, who obsessed a great deal about doing her job correctly, including showing up on time, keeping her coworkers satisfied with her performance, and not drawing any negative attention to herself. As she prepared to go into work, Ruth felt nauseous many mornings and focused on this nausea as a sign of something being wrong. The nausea was actually related to slowing of digestive processes as the amygdala ramped up the FFF response in reaction to her worries about work. Even though there was no physical threat, Ruth's amygdala still produced a defense response to the threat of going to the office. Once Ruth realized that the nausea was due to her FFF response and did not mean she was ill or that any serious danger existed at work, she pushed through the nausea and made it to work on time, rather than calling in to say she was sick. She realized that focusing on that queasy sensation in her stomach and worrying about it was not necessary. She was not ill; she was experiencing the FFF response.

As Ruth got to work and got busy doing her job, the queasiness went away. The amygdala was trying to prepare her for a threat, but when nothing threatening occurred, the FFF response ended. Until Ruth understood what was occurring, the defense response created by the amygdala was getting in her way more than it was helping. If we can help you *reinterpret* your experience of anxiety, and see it for what it is—a part of the defense reaction produced by the amygdala—anxiety will be less likely to continue to fuel your obsessions or compulsions. When you experience anxiety, you often assume that it means that a danger exists. Actually, what it typically means is that the amygdala is creating the defense reaction.

Finally, when it creates the FFF response, the amygdala can have another important impact. When the amygdala is activated, it has the ability to override the cortex and take charge. The truth is, the cortex doesn't work well when the amygdala is producing a strong defense response. Many neural connections go from the amygdala into the cortex, allowing the amygdala to strongly influence the cortex's responding on a variety of levels (LeDoux and Schiller 2009), including directing attention to what the amygdala considers important. Because of its capacity to override other brain processes (LeDoux 2015), the amygdala can influence what we focus on, how accurately we process incoming information, what memories come to mind, and what actions we take. When people have trouble concentrating or find themselves frozen in fear, they may think they're losing control or going crazy. This is what Daniel Goleman (1995) called an "amygdala hijack"—when the amygdala takes charge and the cortex is sidelined.

When your cortex is having difficulty focusing, it does not make sense to you, and you may think about what you are experiencing in your cortex in ways that produce even more anxiety. Because Sophia can't focus on hearing the questions the CEO is asking her, she doesn't just worry about her job interview; she starts worrying that she is losing control of her mind. People with OCD often focus on these types of experiences and obsess about the idea that they may "go crazy." But it's completely normal that you can't focus your attention or organize your thoughts when the amygdala has hijacked your cortex. This does not mean you're going crazy. The brain is simply hardwired to allow the amygdala to take control when it detects danger. Instead of obsessing about these experiences, we will encourage you to learn ways to have more control over the cortex and its effect on your amygdala, using techniques we'll share in part 3.

From the FFF Response to Feelings of Fear and Anxiety

When the amygdala produces the FFF response, we feel many changes in our bodies, and as we process this defense reaction in our brains, we may experience a distressing emotion: a feeling of fear, anxiety, or dread. Let's take a closer look at how the FFF response leads to these emotional reactions. The amygdala does not produce these emotions directly; to feel an emotional reaction, we need the brain to process our experiences in a way that allows us to be aware of our emotional reaction. In the same way that the cortex helps us process what is coming from our sense organs, the cortex provides us with consciousness of our feelings. But these feelings are based on and influenced by the defense reaction that the amygdala creates. This is why people often say that the amygdala is the source of fear and anxiety, although technically it's more complicated than that (LeDoux 2015).

For us to have *emotions*, we have to have consciousness of what is going on in our bodies and brains. That takes a variety of parts of the brain that we're pretty confident you don't want to know the details of. Let's take a simple example—seeing—that illustrates the way the brain works so that you understand why we don't want to say the amygdala creates fear and anxiety. You might be tempted to say that you see with your eyes, but your eyes are only receiving light waves and transferring them into electrical signals that can be processed by your brain. Certain parts of your brain need to receive and process these electrical signals for you to see anything. If these parts of your brain (which happen to be at the back of your head in your occipital lobes) are not working, you will not be able to see. If you have damage to your occipital lobes, due to a gunshot wound or a stroke, *even if your eyes work perfectly*, you won't be able to see. Your occipital lobes are the part of the brain that allows you to see, not your eyes.

Just as your brain processes visual information, parts of your brain process information about your bodily sensations, thoughts, and so on, to produce your experience of emotions. The amygdala is very involved in creating and influencing the physical sensations that contribute to your feeling fear and anxiety, but *feeling emotions* depends on parts of your brain that help you interpret that information so that you know what you are feeling. So, just like seeing things relies on your occipital lobes to interpret the information coming from your eyes, feeling emotions relies on various parts of the brain to identify and interpret these emotions. So while the amygdala is a major contributor in creating aspects of your experience of anxiety, you actually *feel* anxiety and fear because of a variety of brain processes that bring emotions into consciousness. If you want to be neurologically correct, you can consider the amygdala the root of fear and anxiety, but it's not the only structure involved in our conscious experience of these feelings.

Designed to Feel Dread-Full

While the FFF response is the defense response created by the amygdala, remember that the emotional reaction we're experiencing is a whole different process. Often the emotion is a feeling of impending danger or dread, frequently called anxiety. Extremely strong anxiety is better called "panic," which is a terrifying feeling to have. Anxiety is *designed to be distressing* in order to strongly motivate us to do something about the situation. In other words, it's an uncomfortable emotion that demands our attention be focused on a potential danger. Anxiety is a feeling of dread meant to motivate us to do anything to make the feeling stop. It is very often the underlying reason we engage in obsessing, worrying, or carrying out compulsive behaviors. Anxiety can serve to motivate these thoughts and behaviors, and the reduction of anxiety serves to reward compulsive behaviors.

The experience of anxiety means that we *feel* as though something is wrong. This is not a *thought* that something is wrong. We *feel* as though something terrible has happened, is happening, or will happen. The feeling is very real, and impossible to ignore. Often friends or family will try to reassure us that the feeling is not real. Nothing could be further from the truth. When you experience either anxiety or fear, *you are experiencing a very real and very uncomfortable emotional reaction in addition to other changes in your body.* We don't want to minimize the fact that you are experiencing a distressing emotional reaction. But be aware that even when you feel the very real feeling that you are in danger, it is also true that *you may not be in danger at all.* The feeling does not accurately reflect true danger—it simply reflects the amygdala's reacting *as if* there is danger. This is a very significant piece of knowledge to keep in mind.

You can think of it this way: The amygdala is like an alarm system in your body, but the alarm is not signaled by a flashing light on a dashboard or a sound coming from a smoke detector. The signal that's being sent is a *feeling.* Instead of seeing a signal of danger, you directly *feel* as if there is a danger. So when people say you should just ignore the feeling, they don't understand that it's not as easy as ignoring the flashing dashboard light or a beeping smoke detector. This is a feeling—the feeling you would get if you thought a fire was being signaled by a smoke detector—not the sound of the smoke detector itself. The feeling is much harder to ignore because it's a very convincing indication that you are in danger. When obsessions continue to repeat, and you can't let them go, the convincing experience of anxiety is there, making it feel like you need to focus on those obsessions. When you try to resist the compulsion, anxiety builds intolerably, and when you perform the compulsion you feel such relief from the anxiety. Anxiety is fueling so many aspects of OCD. But as real as the feeling of anxiety is, it still can be an erroneous feeling.

When we make this point to our college students, we explain that we can experience anxiety when it is incorrect. A typical example we use is "Are you ever very anxious before an exam, and end up getting a good grade?" Students start to recognize that anxiety and fear are not necessarily predictive of what happens. You can be panicked that you don't know where your car keys are, thinking that you won't be able to find them, you'll be late for work, you'll miss an important meeting, and so on, and be feeling the worst dread and panic. And then you find your keys in your coat pocket. The feelings of fear and anxiety produced by the FFF response do not predict what will happen. They are real feelings but not necessarily accurate predictors.

As you get to know the amygdala, you see how often it may be reacting in error—we may be approaching a piece of vine, not a snake at all. But, it's easy to be fooled into trusting the feelings of fear, anxiety, or dread. Even in the vine situation, when we recognize that the object is *not* a snake and feel some relief, the FFF response does not immediately disappear. The effects of the adrenaline that was released, the increased heart rate, and the muscle tension often remain for a period of time. How are you to interpret that? You're still feeling the effects of the FFF response in your body for several minutes, so it can be difficult to convince yourself that you really are safe.

If we can remind ourselves what anxiety really is—our brain's interpretation of a set of physiological reactions based on centuries of human experience when fighting and fleeing were essential, but which don't always indicate danger—we learn to experience these reactions differently. This is especially important if your obsessive thoughts or worries relate to your physical sensations or health. For example, Malika has trouble breathing when she experiences anxiety, and when her breathing is off, she's afraid to drive because of thoughts that she will pass out behind the wheel. Once she recognized that her breathing issues were part of the FFF response, she realized that she was not going to pass out. Malika found that as soon as she got in the car and began driving, she could stop thinking so much about her breathing and get to her destination, especially if she turned on the radio and listened to an entertaining morning show. She wouldn't have had the courage to ignore her worries about her breathing and get into the car if she hadn't understood the FFF response.

How we interpret the physical sensations of the FFF response is very important. When we experience this defense response, we are feeling the effects of adrenaline, muscle tension, or digestive upset. But we are not experiencing a heart attack or stroke. We are not going crazy or losing control. Also, although we may be experiencing a dreadful emotional reaction, we are not necessarily in danger and should not assume that disaster will soon befall us—even though that's exactly how it feels. Of course, this is more easily said than done, but it does help to remember that many times when you felt this way, everything turned out fine, even though your body was telling you otherwise.

Hopefully, you're learning some things about your own physical and emotional experiences as a result of knowing how they are created in the brain. First, you can see that these experiences are produced by a part of your brain that you don't have direct control over, and that this part of the brain can create changes in your body that lead to physical and emotional responses that occur based on information that may not be in your awareness. Second, you may be able to see these physical and emotional responses in a different way when you recognize that they have the purpose of helping you escape or fight back against danger. And you can also recognize that you shouldn't trust your feelings to indicate that bad things are about to occur.

Before you get the idea that you might be better off without the amygdala, remember that its role is to protect you, even if it's not always right about whether something is truly dangerous. And every once in a while, you do need to run or fight to protect yourself, even though (thankfully) that's no longer typical in the modern world. Remember also that the amygdala produces some positive emotions due to association-based learning. For example, you may have objects in your home that may seem meaningless but that you cherish dearly, like the old wooden-handle potato masher that Catherine associates with Sunday dinners at her grandmother's house. The amygdala also underlies the positive feelings associated with old high school jerseys that many former high school athletes hold on to long after the details of their winning seasons are forgotten.

In summary, you have learned that you have a protector in your brain, the amygdala, that has been a part of the human experience since ancient times. You have learned that the amygdala has a defense response system that can take over both your brain and your body when it feels you are in danger. You have learned that when the defense response is produced in your body, you also have emotional reactions that can be distressing. But you have also learned that the amygdala is not always correct, partly because it makes snap judgments on the basis of limited information. While this method may have protected our ancestors and may occasionally prove very useful to us, at times it does not seem to fit our twenty-first-century world. The more you understand the processes that go on in your amygdala, the more you'll be able to understand that because of the amygdala, anxious feelings can sometimes occur when they are unnecessary. As you learn to be more aware and observant of how your amygdala is reacting, your experience of anxiety may change. In addition, knowing that the amygdala is the root of fear and anxiety in the brain, you can learn methods to calm your amygdala so that it produces the defense response less and you experience less anxiety. If you have less anxiety, you'll have fewer obsessions, and you'll also be more able to resist compulsions. Before addressing methods of calming the amygdala in part 2 of the book, however, we will focus in the next chapter on how the cortex contributes to your OCD.

CHAPTER 3

The Cortex
The Key to Creating Your OCD

You have learned that the amygdala is the root of the anxiety that serves as fuel for your OCD. But many times, especially for someone with OCD, the amygdala is not where the experience of anxiety is initiated—the cortex is. The cortex is very involved in the process of creating and strengthening obsessions and increasing the anxiety that motivates compulsions. The cortex operates in a completely different way than the amygdala, so a clear understanding of the cortex's role is needed. When you understand how the cortex contributes to anxiety, you will have more effective tools for fighting OCD.

Origins of Anxiety in the Cortex

If you deal with obsessions or compulsions, your cortex is probably very involved in activating your amygdala when it is unnecessary. Remember—the cortex cannot create the defense response; only the amygdala can do that. But the cortex can activate the amygdala to create the defense response in two general ways. The first way is through the manner in which the cortex processes sensory information. You will recall that, when the thalamus receives sensory information from the eyes, ears, and other sense organs, it directs this information to the cortex as well as to the amygdala. As the cortex processes information, it can interpret perfectly safe sensations as threatening. The amygdala, which receives the sensory information *before* the cortex does, may not have identified the sensory information as indicating danger, but the cortex can produce thoughts that indicate to the amygdala that the sensory information does indeed pose a danger. In this case, the cortex turns a neutral experience (that wouldn't typically activate the amygdala) into a threat, causing the amygdala to create the defense response.

Let's look at a common example among individuals who have health-related obsessions. If Sheila wakes up with a headache one morning, her amygdala is not likely to interpret the sensations of a headache as a threat. But if Sheila's cortex begins to think about brain tumors and begins to worry that the headache means that she has a cancerous brain tumor, this thought in her cortex may activate the amygdala. And amygdala activation can be worsened if Sheila takes some time to google symptoms of cancerous brain tumors and begins to think that she has other symptoms, such as fatigue and drowsiness. This is an excellent case of the cortex creating thoughts and images that the amygdala will react to. The headache did not activate the amygdala directly: The cortex's *interpretation* of the headache is what caused the amygdala to react. The power of interpretations to activate the amygdala is illustrated in figure 3.

The second way the cortex can activate the amygdala does not require the cortex to be processing any sensations. For example, as Tony is riding the train to work one morning, he is not experiencing any danger. The train is running on time; it's a beautiful, crisp fall morning; and he's enjoying a delicious, warm cup of coffee. None of his sensory experiences indicate danger. But Tony begins thinking about his girlfriend, and it occurs to him that she has not texted him this morning. He begins to think about their relationship. He worries he has not been attentive enough and wonders whether his girlfriend is going to break up with him. He suddenly feels anxious and feels the need to text her, but then he worries that he may wake her and cause her to be irritated at him. These thoughts and images have not been prompted by any sensory information at all—Tony's cortex has simply created them. When distressing thoughts or worries are produced in the cortex, the amygdala can be activated to produce a defense response, even when the person has not seen, heard, or felt anything dangerous.

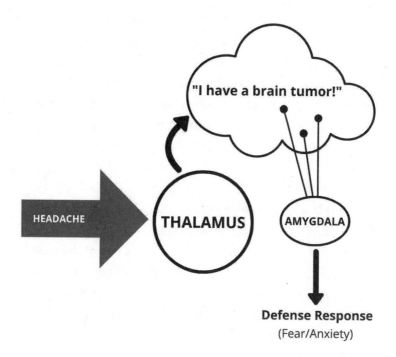

Figure 3

This shows you how the cortex can cause the amygdala to become activated, and how this can lead a person to experience anxiety. Various thoughts go through our minds all the time. Many thoughts produced in the cortex are not necessarily informative, correct, or worth your attention. But, especially when you have OCD, you're more likely to take troubling thoughts seriously, and when the amygdala reacts to these thoughts, resulting in anxiety, the concern you have about the thoughts seems to be validated. This is a mistake, because the amygdala's tendency to respond to thoughts as though they are dangerous is not accurate. Thoughts themselves are not dangerous, whether or not they provoke activation in the amygdala.

This is a key point about the thoughts in the cortex. People with OCD have a tendency to think that some thoughts are dangerous. Sheila's thought about brain cancer and Tony's thought about problems in his relationship are not dangerous thoughts. They do not pose a danger in any way: by themselves, thoughts cannot cause cancer or make problems form in the relationship. They are simply thoughts. What they *can* do is activate the amygdala, causing the defense response to occur. This reaction can add to a person's concern about thoughts being dangerous. Now Sheila and Tony are having certain thoughts and feeling a reaction in the body at the same time. The amygdala's reaction can cause them to experience anxiety and contribute to their concerns about their thoughts.

Cognitive Fusion

We need to explain this process of taking thoughts too seriously because it frequently contributes to cortex-based anxiety. *Cognitive fusion* is believing in the absolute truth of mere thoughts. This is one of the most basic cortex-related processes that makes OCD occur. It's one thing when the amygdala makes the error of reacting as though thoughts are reflecting reality; it's another when the *cortex* also makes the error of rigidly believing that thoughts it experiences reflect reality (cognitive fusion). In other words, *you* believe that a thought is true, or could be true. When this happens, a great deal of unnecessary anxiety frequently results. In the examples above, both Sheila and Tony are falling victim to cognitive fusion. They are assuming that a thought is reflecting reality. People who experience obsessive thinking frequently take mere thoughts too seriously, and the belief in the validity of mere thoughts is part of what leads those thoughts to be turned into obsessions.

When Sheila *believes* her interpretation of her headache—the *thought* that she could have a cancerous brain tumor—it is difficult for her to stop thinking about it. As she thinks about it, her amygdala is activated and produces the defense response. Sheila feels anxiety, and this feeling leads her to believe that there is some danger, and of course, she is likely to interpret the feeling as validating her thought: *If I feel like I am in danger, I must have something terrible occurring in my head.* As you can see, not only thoughts, but a feeling, such as anxiety, can be taken too seriously. When Tony worries and feels anxiety about his relationship, he may think that his *thoughts* and his *feelings* of anxiety mean that the relationship is in danger. You can see how easily this can occur, especially if one is not aware of the relationship between the cortex and amygdala in creating anxiety. When a simple thought activates the amygdala, resulting in the feeling of anxiety, the average person is likely to take the thought more seriously. *I feel anxious, so there must be something I should be concerned about in this thought.* Hopefully, you're beginning to see how you can experience anxiety even when no true danger exists.

Cognitive fusion occurs quite often because we tend to overvalue the processes that are occurring in the cortex. Humans have a tendency to overestimate the accuracy of our thoughts and fail to remember how many times we have assumed something in error. We think of ourselves as logical, intelligent people, governed by reason, but that is not an accurate description of how our brains really work. We often misinterpret situations and don't always have access to accurate information about what is occurring. Nevertheless, we tend to operate as though our perceptions and interpretations are valid. The cortex has a tendency to operate as if it possesses the real meaning of every thought, emotion, or physical sensation. In reality, humans are prone to erroneous, unrealistic, illogical, and even random thoughts that don't make much sense. You do not need to take every thought that pops into your mind seriously.

Do you have difficulty allowing thoughts or emotions to pass through your mind without giving them a great deal of undue attention or analysis? In chapter 10, we'll provide strategies you can use to "defuse" your thoughts.

Anxiety caused by the cortex can be initiated with or without the interpretation of sensory information. We'll explore both of these methods separately below.

Cortex-Based Anxiety *with* Sensory Information

Many times, you may be in a situation that's perfectly safe, but your cortex is responding to the sensory information coming in as though it is dangerous or problematic. Your cortex does not just receive information; it also *interprets* the information, and this interpretation can be faulty. Our cortex's ability to go beyond the information and to apply additional meaning to a situation can be amazingly helpful or disastrously incorrect. To understand this process, we need to take a closer look at the frontal lobes of the human cortex. As we noted in chapter 1, the frontal lobes have given us the ability to anticipate and imagine things we've never seen. They help us contemplate future events and consider potential consequences. This also means that the frontal lobes are one source of anticipatory thoughts that can contribute to obsessions.

Having anticipatory thoughts is not always a problem. In fact, the nature of human intelligence is to try to predict what will happen. Without the ability to anticipate problems and to find solutions, we could have never learned to be successful at hunting, planting and harvesting food, caring for our children, building skyscrapers, or designing safe automobiles. But our ability to produce anticipatory thoughts also has a serious disadvantage: we suffer through many situations that never actually occur.

This human capacity to anticipate future events means that, from time to time, everyone will have worries or occasional distressing thoughts that are difficult to get out of their heads. But when you have OCD, you're often dominated by a variety of distressing thoughts for hours each day. They can make it difficult to get through the average day. In fact, some of the most intelligent and creative people have brains that seem to use all their impressive intellectual abilities to produce very compelling, distressing thoughts that seem impossible to stop or disregard. Your intelligence and creativity can actually work *against* you, because they help you come up with even more distressing thoughts and conjure up convincing reasoning for having concerns about them.

When you're being dominated by obsessive thoughts, the frontal lobes are repeatedly producing thoughts and images that cause the amygdala to be activated. Studies have shown that people with OCD show more activation in frontal lobe areas as well as the amygdala compared to those without OCD (Thorsen et al. 2018). The way your cortex interprets events will influence whether you're interpreting a situation as safe or dangerous. In the OCD brain, many safe situations are turned into ones that activate the amygdala, all because of the interpretations provided by the cortex, and especially by the frontal lobe's ability to anticipate. Whether this is due to certain learning experiences, specific inherited neurological processes, or a combination of both, the circuitry in the cortex can respond in ways that promote worry, pessimism, and other negative interpretations of sensory experiences that activate the amygdala.

Let's consider Marjorie, who is giving Teddy a bath. Teddy is having a wonderful time, and at one point puts his face into the water to blow some bubbles. Even though Teddy is giggling, for some reason Marjorie thinks about how it would look if Teddy were drowning. This thought terrifies her. She then thinks about how it would look if she held Teddy's head under the water. She thinks to herself how easily she could drown a toddler like Teddy. This thought really distresses her, and she begins to wonder whether she could potentially harm Teddy. She feels anxious about bathing Teddy after that and asks her husband to bathe him. Her anxiety causes her to avoid even seeing Teddy in the bath.

As we saw in figure 3, the person's interpretation is what leads to anxiety. Marjorie's *interpretation* of Teddy blowing bubbles in the water caused her amygdala to produce anxiety. She saw her toddler blowing bubbles and interpreted the situation as similar to how it would look if her toddler were drowning. Her amygdala was not reacting to the child blowing bubbles—it was her thought of her own child drowning that activated the amygdala. Many other ideas could have come into Marjorie's mind as she bathed her child. She could have thought that she should grab her camera, or she could have told Teddy he was a silly boy, or not to drink bathwater. But Marjorie's particular thought activated her amygdala. As a result, she experienced anxiety. Then she focused on the thought and continued to imagine ways it could happen, including the idea that she herself could drown her child. Her anxiety is clearly cortex-based because it came from interpretations produced in her cortex, even though it was related to events occurring in the environment.

Another important aspect of Marjorie's reaction is her interpretation of this thought. Notice that Marjorie treats her thought as though it means something. She is showing cognitive fusion. She believes in the truth of mere thoughts when she assumes that her thought of how she could drown her child indicates she may actually harm her child. She doesn't realize how normal it is for a parent to have a random thought about a drowning child. Instead of simply thinking, *What a weird thought to have!* she started worrying about what the thought meant.

Marjorie's thoughts about her child drowning are examples of *intrusive thoughts*: thoughts that unintentionally come into a person's mind. They are very common in all individuals, which is something that most people with OCD do not recognize. We have long known that a large majority of people (84%) report unwanted intrusive thoughts similar to Marjorie's thoughts (Rachman and de Silva 1978). Since then, numerous studies have found that most individuals (80% to 90%) experience unwanted intrusive thoughts that are very similar in form and content to the ones people with OCD experience (Radomsky et al. 2014). Most people have thoughts randomly pop into their heads about harming others, about disastrous events occurring, about ways to commit suicide, and about sexual behaviors they would never perform (Radomsky et al. 2014). The difference is that people with OCD are more likely to focus on an unwanted intrusive thought and evaluate it more negatively (Gibbs 1996). In summary, the thoughts are not the problem; the *focus* on the thoughts is the problem.

Returning to Marjorie, we need to remember that her thought about her child drowning didn't necessarily mean anything. But she took it too seriously, due to cognitive fusion, continued to think about how it could happen, and then had the frightening thought, *I could harm my child!* Notice that the fact that this thought was frightening is actually a good sign—Marjorie is obviously concerned about her child's safety and does not seem likely to harm her child. If she thinks of her child being in danger, or if she thinks of harming her child, she experiences distress. She is a protective mother. But this is not how she interprets her thought or her anxiety. She interprets her thought as a prediction of what she might do, and when her amygdala is activated, she interprets her experience of anxiety as an indication of true danger. This manner of reacting to thoughts produced in the cortex is what fuels OCD. People with OCD do not only take their thoughts too seriously (*My child could drown!*); when the thoughts activate the amygdala and as a result they experience anxiety, they become more convinced of the dangerousness of their thoughts and come up with additional interpretations (*I could drown my child!*) that further activate the amygdala. When thoughts activate the amygdala, producing the defense response and leading to the feeling of anxiety, they interpret the anxiety to confirm that a danger exists, when all that has occurred is a mere thought.

When you have OCD, you need to learn to resist cognitive fusion. Try to remember that everyone has the same kinds of thoughts you do. When we ask a classroom full of college students how many have a random thought like *If I turned my car in front of that semi-truck, I could kill myself* as they are driving down the freeway, nearly all students raise their hands. This often stuns those with OCD, who believe that having such a thought is extremely significant, not a common occurrence. In actuality, if you take the thought too seriously (cognitive fusion) and continue to think about it, you're setting the stage for obsessive thinking to take root. When you continue to think a thought over and over, you're making it stronger and stronger. "Survival of the busiest" is the law of the cortex. The busiest circuits are strengthened, and unused circuitry weakens.

Cortex-Based Anxiety *Without* Sensory Information

Sometimes anxiety begins with thoughts in the cortex that occur without any sensory information at all. This means that a person is in a safe situation and is not seeing or hearing anything that prompts thoughts about danger, but nonetheless the cortex activates the amygdala. Two categories of this type of anxiety often occur in OCD: thought-based and imagery-based.

Thought-based anxiety arises in the left hemisphere (side) of the brain, which is responsible for language and logical reasoning. This hemisphere is also responsible for both worry and rumination (Engels et al. 2007). *Worry* is the process of thinking of potential negative outcomes for a situation. In chapter 4 we'll explain how worry contributes to OCD. (You can also read the bonus chapter "The Healthy Use of Worry," available at http://www.newharbinger.com/47186, to learn how to use worry in healthy ways. Please see the very back of this book for more details about this and other online accessories.) *Rumination* is a style of thinking that involves repetitively mulling over problems, relationships, or possible conflicts. When you're ruminating, you're intensely focused on repeatedly considering the details and possible causes and effects of situations (Nolen-Hoeksema 2000). Many people mistakenly believe that worry or rumination is helpful and that logically thinking about something for a great deal of time is likely to lead to a solution to a problem. The truth is, more thinking about a situation does not necessarily mean that you'll improve the situation. In fact, worry increases stress in the body (Virkuil et al. 2009), and rumination has been shown to lead to depression (Nolen-Hoeksema 2000).

The right hemisphere of the brain offers another way to activate the amygdala, even in the absence of any dangerous situations. Unlike the left hemisphere, the right hemisphere is not analytical and logical and does not involve language; the right hemisphere is nonverbal and processes things in more holistic, integrated ways. It provides us with visual images, imagination, daydreams, and intuition. It helps us see patterns, recognize faces, and identify and express emotions. Because of these capacities, the right hemisphere contributes to anxiety through visualization and imagination.

Marjorie's image of drowning her child came from the right hemisphere of her cortex. When you visually imagine violent or sexual scenes you find disturbing, you are using your right hemisphere. When you imagine the critical, disappointed look on your partner's face when she finds out you forgot to stop at the store, your right hemisphere is involved. If you're particularly good at using your imagination, you can expect your amygdala to be activated frequently by stress-producing or frightening images you create in your right hemisphere. In fact, amygdala responding seems to be worse when people are focusing on distressing images than when they are worrying (Freeston, Dugas, and Ladouceur 1996).

You can see that the cortex has the ability to use thoughts, interpretations, worry, rumination, and frightening images to activate the amygdala. People usually find it easier to recognize the cortex's role in activating the amygdala because we are more aware of what happens in the cortex. We are more able to observe the language of thoughts and images produced in the cortex than we are to observe or understand the amygdala. Some parts of the cortex are also more directly under our control than the amygdala is, which means we are more able to influence the cortex's processes than the amygdala's processes. In fact, cognitive therapy is an approach that focuses on changing cortex-created thoughts and images in order to modify a person's feelings. Aaron Beck (1976) and Albert Ellis (Ellis and Doyle 2016) have focused on helping people change their cognitions (i.e., thoughts) and found beneficial results for a variety of disorders. The cognitive approach gives us some specific ways to combat OCD in the cortex, and studies have found that cognitive therapy can result in significant improvement in OCD symptoms (Whittal, Thordarson, and McLean 2005). Although it's not always easy, changing what is happening in your cortex often helps you avoid activating the amygdala. We will explain ways to use cognitive therapy as well as other approaches to change your cortex in part 3 of this book.

The Amygdala: Part of the Cortex Pathway to Anxiety

Remember that cortex-based anxiety still requires the involvement of the amygdala, even though it is initiated in the cortex. (See figure 2 in chapter 2 for a reminder.) The cortex does not have a way to produce the defense response—only the amygdala does. When information about threats or danger in the cortex is detected by the amygdala, the amygdala produces the defense response that leads to our experience of fear and anxiety. If you could examine the connections in your brain, you'd see many connections from your amygdala to the cortex (LeDoux 2002), giving it a great deal of influence in the cortex. Connections from the cortex to the amygdala are much fewer. This means that the amygdala is constantly monitoring and influencing what is occurring in the cortex, while the cortex has less oversight or influence over the amygdala. It's difficult to intentionally control our emotions with our cortex: often our emotional reactions seem better at controlling us!

This is an important fact about the relationship between the cortex and the amygdala. Because the amygdala has more connections to monitor and influence the cortex, at times the amygdala has more influence and control over our mood and our feelings than the cortex does. This is true even though, at times, the cortex may have more accurate information. For example, the cortex is able to recognize the curved object on the ground as a vine rather than a snake better than the amygdala can, because the cortex has more detailed visual information than the amygdala does. Luckily, even though the amygdala causes you to panic and jump away from that vine, you can fairly quickly recover after the amygdala also takes into account more detailed information from the cortex.

The important point to take away about the complex relationship between the cortex and amygdala is that thoughts and images that are produced in the cortex have a strong influence over the amygdala's reactions. We have used the image of the amygdala watching "Cortex Television" to illustrate the type of response the amygdala has to the cortex. Because of the many connections the amygdala has to the cortex, you can think of the amygdala as observing whatever is occurring in your cortex. Distressing or threatening thoughts or images in your cortex can activate the amygdala to produce the defense response, leading to fear and anxiety. Your cortex can also produce reassuring images or thoughts to help keep the amygdala calm, or help return it to a calm state once it has activated the anxiety response. You should note, however, that once the amygdala has activated the defense response, it can take a while for the body to recover. Calm images or thoughts from the cortex might help, but they can't cancel physiological processes once they have been created by the amygdala.

You may be surprised to learn that the amygdala responds to something you are *imagining* in your cortex much like it responds to something that is actually occurring. Ideas and images that we conjure up in our cortex are capable of activating the amygdala even though they are not real. Neuroscientists don't yet know whether the amygdala has a way of separating thoughts and images that are grounded in reality from those that are imagined or assumed (LeDoux 2015). Have you ever experienced a terrible feeling of dread and panic when you were just imagining some disastrous event? That feeling comes from the amygdala reacting to your imagination as if it were reality.

You may wonder why the amygdala would be so easily activated by information it is monitoring in the cortex. The truth is, it can be beneficial for the amygdala to react to thoughts or images in the cortex as if they were real. Let's look at the example of sixty-five-year-old Charlotte, who is at home one evening when she hears the familiar sound of someone coming in the back door. She hears the sound of the back door every night when her husband comes home, so her amygdala (which processes the sound first) does not respond to it as a signal of danger. But Charlotte knows *in her cortex* that her husband is away on a fishing trip and that no one should be coming in the back door at this time. Her cortex puts together this information and imagines a stranger entering her home, and it then produces the thought that she may be in danger. These thoughts and images in her cortex prompt Charlotte's amygdala to become activated. Her heart starts pounding and she stops what she is doing. She becomes hypervigilant to other sounds and focuses on getting herself out the front door to safety. If there is an intruder, these reactions could save her life.

Note that in this example, Charlotte's amygdala is not activated by the sounds of the door opening, because they are common, familiar sounds. The amygdala does not have the ability to interpret the meaning of the sounds in the way the cortex can. It is not designed for the kind of detailed analysis the cortex can provide. The amygdala is activated only when it responds to Charlotte's cortex producing *thoughts and images* that indicate an intruder may have entered the house. Responding to information from the cortex allows the amygdala to guard against dangers it may not recognize. The amygdala relies on the cortex to provide it with additional information and monitors the cortex for such information. This shows how the amygdala's ability to observe and respond to the processing of information in the cortex can be very helpful in some circumstances.

Unfortunately, at times, the amygdala's reliance on the cortex can lead to unnecessary and unfounded anxiety. Let's consider another situation that could occur with Charlotte. In this example, Charlotte is once again alone at home and her husband is away on a weekend fishing trip.

She doesn't hear anything unusual, and it's a typical night as she prepares to go to bed. She feels uneasy when her husband is away, and as she lies in bed listening to the quiet night, she imagines that someone could break into her house. She imagines an intruder walking around in her yard carrying a knife, and imagines that he could use the knife to slice open the screen in her bedroom window. Charlotte's amygdala is going to respond to these images. Even though no evidence suggests she's in danger, her amygdala still responds to activity in her cortex by initiating a defense response. She will feel her heart pound, be alert for any sound, and feel a sense of fear. She may begin to think she should hide or seek help, even though, at the same time, she may realize that there's no evidence of any danger. Her emotions are coming from the amygdala responding to what is happening in the cortex as indicating danger. In this situation, the amygdala relying on the cortex is not helpful.

Knowing that your amygdala can respond to thoughts and images of danger as if they reflect actual danger allows you to recognize that what you think about and imagine in your cortex definitely affects your anxiety level. From the perspective of the amygdala, thoughts and images in the cortex may call for a response, even if the amygdala itself does not detect danger from the sensory information it received. In some cases, the amygdala's decision to rely on the cortex may save a life, and in some cases, it may result in unjustified anxiety. In either case, once the amygdala gets involved, you will experience the physical sensations associated with the defense response and likely experience anxiety. And because you'll feel the sensation of dread and your body reacting as if you are in danger, you may have difficulty knowing whether or not you are safe.

You can see how the cortex's tendency toward cognitive fusion can lead you to trust the accuracy of thoughts or feelings, and this can further activate the amygdala. You can also see how unwanted intrusive thoughts can lead to problems if you take them too seriously. This is why we often warn people, "Don't scare your amygdala!" if they have a tendency to focus on frightening thoughts or images in the cortex.

Fortunately, we can interrupt and change cortex-based thoughts and images that may activate the amygdala when it isn't necessary. With practice, you can use a variety of different strategies to rewire your cortex to be less likely to activate your amygdala. The first step is to become aware of (and possibly write down) the thoughts and images that your cortex is producing that may be activating your amygdala, leading to anxiety, and to *recognize that they are just thoughts and images*. This will help give you the needed motivation to change them. But before you're ready to make any changes in the cortex, you need to have a better understanding of two more cortex-based problems that contribute to OCD: worry and obsession. We'll explain both of these in the next chapter.

CHAPTER 4

The Roots of Worry and Obsession

When we humans express pride in all the amazing capabilities of our brains, we are focused specifically on the cortex. The cortex is the source of the remarkable human abilities to imagine, plan, and create what has never been seen. But you are learning that these abilities come at a cost. Because we can imagine what may never come to pass, the cortex is also often the source of unnecessary anxiety. In this chapter, we'll discuss processes in the cortex that produce worries and obsessions as well as the processes in the brain that sustain them. We have a tendency to overestimate the importance and accuracy of the thoughts produced in the cortex. If you take some time to understand what the cortex is actually doing, and recognize its limits, you can learn how to protect yourself from the negative effects of worry and obsessions.

As we have noted, because the human cortex can imagine and predict future events and their consequences, we experience anticipation, a thought process in which we form expectations about what might occur. Anticipation requires the cortex to begin preparing for a future event by considering or visualizing it. Anticipation requires a variety of areas in your cortex to work together, and researchers are still identifying those areas (Andrzejewski, Greenberg, and Carlson 2019; Grupe and Nitschke 2013), but much of the circuitry is in your frontal lobes (the part of the cortex behind the forehead). This part of your brain is also where many of the neural processes that create OCD are operating.

Anticipation is both a blessing and a curse. We can anticipate in positive ways and feel eager and excited, and we can anticipate in negative ways, expecting and imagining undesirable or even disastrous events. Negative anticipation can lead to a great deal of stress, especially in those who have OCD. Much unnecessary suffering occurs, because the experience of anticipation is often more distressing than the event that's being anticipated. In many cases, the thoughts and images that people reflect upon before an upcoming situation, such as a threatening storm, a potential confrontation, or a job interview, are much worse than the actual situation turns out to be.

The anticipation of negative situations is an essential problem in OCD. As you have learned, the cortex's ability to create thoughts and produce images often results in the amygdala initiating the defense response, even when the person is in a safe situation. Anticipation creates threatening thoughts and images that can significantly increase anxiety, and in OCD, it can focus on a variety of themes, such as contamination, perfectionism, religion, sexuality, causing harm, or relationship concerns. Do you frequently use your ability to anticipate in such a way that it produces anxiety? Does your cortex often turn a completely safe situation or a peaceful day into an anxiety-filled one?

When you consider an upcoming situation, or even a potential situation that may never occur, it doesn't matter whether the situation is a positive or negative one. Take the example of a young man considering whether to ask someone out on a date. His cortex will help him imagine various ways that the situation could play out. He could imagine positive responses, and he could imagine rejection. If he anticipates the latter, the thoughts and images in his cortex are going to activate his amygdala, and he is likely to experience anxiety. Note that, due to negative anticipation, your amygdala may activate a defense response and you may experience anxiety even if the events you are imagining *never occur*. And even if a specific event you anticipate does come to pass, you may dwell on it and feel stress much more than you need to before it even happens. Instead of experiencing the situation just once, you experience it—often in a very negative way—repeatedly before it ever occurs. How many of the statements below apply to you?

If there is a potential for danger or illness, I feel like I must consider it.

When I have trouble finding something, I immediately feel like I'll never find it.

I like to have a plan made for possible difficulties, just so that I know what I'll do.

I think a lot about things that people might say that would upset me.

I can be worried sick about something months before it ever occurs.

I often spend my time thinking of solutions for problems that never occur.

If I have to perform or speak in front of a group, I can't stop thinking about it.

I can almost always think of several ways that a situation could turn out badly for me.

When I know something could potentially go wrong, it is constantly on my mind.

If I detect a potential conflict, I spend a lot of time considering it.

A tendency to anticipate negative events is very common in people with certain kinds of OCD, like those focused on causing harm or fear of being contaminated. If you have a strong tendency to anticipate negative events, those thoughts activate your amygdala and create more anxiety in your life than is necessary. Keep in mind that while everyone experiences difficulties in their lives, you shouldn't have to live through these events repeatedly in your cortex when nothing negative is occurring. For example, you could suffer for months through thoughts that you may unknowingly feed your toddler daughter something she's allergic to, when in fact it never happens. Returning to our Cortex Television analogy, it's like constantly watching the Anxiety Channel when there are many other channels you could watch.

Note that, as a certain situation approaches, both anticipation and anxiety will tend to increase, but the increase in anxiety is not predictive of what will happen in the situation; it is simply reflecting the increase in anticipation. We have all had the experience of being very anxious before a test that we end up doing well on. Anticipation is not an accurate predictor. Before an event occurs, your cortex is simply doing more anticipation, and the amygdala reacts with more anxiety. For example, if you're waiting your turn to speak in front of a group of people, your anxiety is likely to go up as the moment for you to speak comes closer. The worst moments are often in the minutes before you speak. But the anxiety goes down after you begin speaking and focus on what you are saying. Some people think, *If I am this nervous before I speak, I'll be even more nervous after I start talking.* But that's not the way anticipation affects the amygdala. We will be identifying strategies to help you modify your anticipatory thought processes in part 3.

The Roots of Worry

Because of our ability to anticipate, humans developed circuitry in our frontal lobes specifically designed to use anticipatory thought processes to create a kind of thinking known as worry. *Worry*, as we noted in chapter 3, is the process of thinking of potential negative outcomes that have not yet occurred. It is different from simple negative anticipation because it tends to be a chain of thoughts, elaborated on in a fairly logical way, and to be a deliberate process. Based on our ability to anticipate, humans can be very focused on worrying, repeatedly thinking about potential future threats, while our pets don't suffer from these kind of thoughts. Humans worry repeatedly about many things that never happen, putting the amygdala through some distressing, but completely imagined, situations. We could save ourselves a lot of anxiety if we took a break from worrying and focused on living in the present moment the way our dogs do.

You can think of worrying as being on the Worry Channel and watching a scary movie (or a series of scary movies) in your cortex. Your amygdala is going to react to these movies and produce anxiety. A woman can be thinking about the possibility of having cancer, for example, and imagining what it would be like to go through chemotherapy, when all that provoked the worry was an upcoming mammogram appointment. Worry takes us away from the present moment and into thoughts about potential negative outcomes, presenting us with scary thoughts and images that may never occur. Why do we do this to ourselves?

Why We Worry

Researchers have found that we seem to get some relief from worrying. Why? One reason is that worrying, which is a thought-based process, keeps us from focusing on imagery that is even more anxiety provoking (Freeston, Dugas, and Ladouceur 1996; Hirsch et al. 2012). In thinking about a situation, we focus on ideas, directing attention away from images, and ideas tend not to produce as much anxiety as images do (Vrana, Cuthbert, and Lang 1986). If a frightening image comes into your cortex and you begin worrying, you are producing a series of thoughts in your left frontal lobes and focusing less on the images coming from the right side of your brain. If we go back to the movie analogy, the thought-based process of worrying is rather like leaving the room when a distressing, violent movie is on television and going into the next room. Now you can't *see* the violence, even though you still are hearing the movie. If you carefully observe your process of worry, you may recognize that you're switching from images to *thinking* about the situation and analyzing it—and this produces a bit of relief, but unfortunately keeps your cortex on the same topic. The amygdala still reacts by producing a defense response, though somewhat less than it does to frightening images. Worry does result in an increase in heart rate (Virkuil et al. 2009), so it clearly produces negative physiological reactions.

Many people mistakenly believe that worrying is a helpful process, when very often it is not. Worry not only increases physiological arousal, but it also *sustains* anxiety because it's more focused on considering negative outcomes than on planning a solution. Worry may lead to planning and problem solving designed to minimize or prevent future difficulties, but worry is not the same as problem solving. For one thing, worry does not stay focused on planning, but tends to keep returning to thinking about potential negative outcomes. As a result, worry can be quite draining. As nineteenth-century writer John Lubbock once said, "A day of worry is more exhausting than a week of work."

The Source of Worry in the Cortex

Worry is the result of multiple brain systems, with the frontal lobes playing an extremely important role (Paulesu et al. 2010). The *prefrontal* portion of these lobes, just behind our eyes and forehead, carries out many thought processes, including anticipating outcomes, both good and bad. Another part of the frontal lobes involved in worry is the *anterior cingulate cortex*, which plays a role in making decisions about how to respond in future situations (Paulesu et al. 2010). The anterior cingulate cortex is an older part of the frontal cortex, near the center of the brain, that has also been identified as important when it comes to OCD (Paul et al. 2018). Recent studies have found structural and metabolic abnormalities in the anterior cingulate cortex of people with OCD (de Salles Andrade et al. 2019), suggesting this circuitry could be involved in producing OCD symptoms. The anterior cingulate cortex serves as a bridge between the amygdala and the prefrontal cortex and helps us process our emotional reactions in the cortex (Silton et al. 2011). Instead of smoothly relaying information back and forth between the cortex and amygdala as it should, for some of us it may get stuck on certain ideas or images. The information flowing back and forth gets caught in a repetitive loop. We aren't sure why, and some evidence suggests abnormal processing of information in the basal ganglia may be involved (Welter et al. 2011).

Applying this to worry, we may be seeing a problem with the anterior cingulate cortex when people get preoccupied with thinking repeatedly about potential problems and are not able to let go of worry. We call this a "faulty worry circuit" in the frontal cortex, and it's just like a computer getting caught in a loop and not being able to go on to the next step. When the anterior cingulate cortex gets hung up like this, the prefrontal cortex has trouble shifting away from thinking about potential threats and going on to problem solving and planning.

If worry is a central problem for you, you might find yourself lost in the process of considering "What if…?" People who get stuck in their worry circuits often think about a potentially threatening situation in depth and repeatedly consider various frightening possibilities, without being able to move on to planning a solution or even thinking other thoughts. If you find yourself agreeing with many of the statements below, you're most likely vulnerable to getting stuck on the Worry Channel. (If so, you will find it helpful to read the bonus chapter "The Healthy Use of Worry," available at http://www.newharbinger.com/47186.)

I'm very good at imagining all kinds of things that could go wrong in specific situations.

When I solve one problem, I tend to find another problem that I need to think about.

I frequently prepare myself for negative events that I fear will occur, but they seldom do occur.

I know I tend to worry about trivial things.

I don't trust people to follow through on what they say they will do.

I sometimes worry that my symptoms are the result of some medical illness that hasn't been diagnosed yet.

When I'm busy at work or with other activities, I don't have as much anxiety.

Even when things are going well, I seem to think about things that could go wrong.

If there's even a small possibility that something could go wrong, that's what I tend to dwell on.

I have trouble falling asleep because of the things I worry about.

I sometimes feel that if I don't worry about specific situations, something will surely go wrong.

I'd rather expect the worst so that I am not disappointed.

The Roots of Obsession

Obsessions are "recurrent and persistent thoughts, urges, or images that are experienced…as intrusive and unwanted, and that in most individuals cause marked anxiety or distress" (American Psychiatric Association 2013, 237). What makes a thought, image, or urge into an obsession is that it keeps coming back, even though the person does not want it and finds it distressing. While worrying is something that we generally realize we're doing, obsessions happen to us and are experienced as intrusive and not under our control. The feeling of a lack of control of the obsessions is one of their most difficult aspects. They interfere with life, appearing at unwelcome times or taking too much time from your day. Some people's brains seem more vulnerable to obsessing, which requires the ability to stay focused on and keep returning to a particular thought. This focus is a useful ability in some circumstances because humans often need to concentrate on a situation and continuously consider details of it in order to accomplish something or solve a problem. But when one becomes stuck thinking too much about something when such sustained focus is not needed, quality of life is often affected.

A second aspect of obsessions is that they cause the person anxiety and distress. They cause distress in a way that is different from worry, because the thought, image, or urge is upsetting in and of itself, whereas in worry, the possible outcome that is being considered is what leads to distress, not the thought itself. The distressing nature of obsessions is often due to *cognitive fusion*, which you may recall is the tendency to take thoughts, images, or urges as accurate indicators of reality. When you assume it reflects reality, a thought, image, or urge is more likely to trouble you and cause the amygdala to become activated. Typically, a person wants to escape from the obsession, but it keeps coming back. Obsessions can focus on a variety of different topics, including contamination, violence, harm, illness, religion, and sexuality, but they all share the characteristic of being repetitive and persistent. What makes a thought, image, or urge repetitive and persistent? In order to explain, we need to look more deeply at the way the cortex operates and examine the neurology of thoughts in the cortex.

Brain Circuitry

If you suffer from OCD, you may take some comfort in simply knowing where your various symptoms are coming from in your brain, because suddenly some of your reactions will make much more sense. You may assume that brain circuitry operates according to a logical process, but unfortunately it does not. The brain reacts according to connections it has made, rather than logic, and knowing this can help you make sense of what's happening in your life. In addition, once you understand the parts of the brain involved in creating worries, obsessions, and anxiety, you know more about how to change the way these parts of the brain respond. In order to change obsessive thinking in the brain, we will help you make changes in the brain's circuitry.

The brain is made up of billions of cells called *neurons*, or *nerve cells*, that are connected to each other. These connected neurons form circuitry that holds your memories; produces your thoughts, images, and urges; and initiates all your actions. The specific connections, or circuitry, between these neurons can be modified because your brain has *neuroplasticity*, the ability to change itself and its responses. On the basis of your experiences, the neurons in your brain change their connections, their structures, and their patterns of responding. When your teacher repeats 5 x 3 = 15, neurons modify the circuitry in your brain to make a connection between the neural circuitry that stores 5 x 3 and the circuitry that stores 15. When you see the golden arches of a McDonalds just before you get some French fries, your brain makes a connection between the neurons that store the images of the golden arches and the neurons that store the taste of warm, salty French fries. Because we understand how neurons make connections, we can help you learn strategies that will allow you to rewire the circuitry in your cortex that leads to obsessions, compulsions, and anxiety.

Neurons are composed of three basic parts (illustrated in figure 4): the cell body, dendrites, and axons. The *cell body* contains the machinery of the cell, including the various parts that sustain and build the cell. Coming out of the cell body are *dendrites*, which look like branches of a tree. Dendrites are an essential part of the communication system between neurons. They reach out to other neurons to receive messages from these neurons. The messages typically travel down the *axons* of other neurons to the *axon terminal*, the end of the axon. Axon terminals and dendrites don't actually touch each other to communicate. Axon terminals communicate by releasing chemicals into the space near the dendrites. These chemicals, called *neurotransmitters*, can cross the space and connect with the dendrite in order to relay a message to the next neuron.

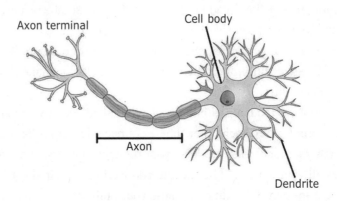

Axon terminal

Cell body

Axon

Dendrite

Figure 4

The space between the end of an axon (i.e., axon terminal) and the end of a dendrite is where the communication process occurs; this space is known as a *synapse* (illustrated in figure 5). Inside the axon terminal, tiny sacs hold neurotransmitters in preparation for sending their chemical messages across the synapse. Some neurotransmitters excite the next neuron, encouraging it to send the message on to other neurons, and some neurotransmitters inhibit the neuron or quiet it. Examples of neurotransmitters are adrenaline, serotonin, and dopamine.

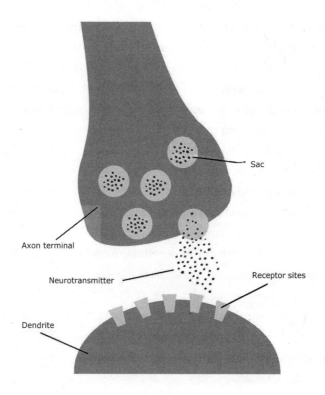

Figure 5

Neurotransmitters are called "chemical messengers" because when they cross the synaptic space, it's as if they are carrying a message to the next neuron. Neurotransmitters connect to *receptor sites* on the dendrite of the next neuron and have an effect similar to putting a key in a lock. If enough neurotransmitter excites the next neuron, it can cause that next neuron to react by firing. *Firing* is when a positive electrical charge travels from the dendrites through the cell body and all the way to the axon terminal of the neuron. This causes the axon terminal to release neurotransmitters from the sacs, transmitting the message to yet another neuron, passing the message on.

Neurons operate on the basis of chemical messages traveling *between* neurons and electrical charges traveling *within* a neuron. All the sensations that you experience, from the sight of these words on this page to the sounds of your dog barking, are processed in your brain by neurons. The sensations that you experience, such as light waves entering your eye or the vibrations in the air that impact your eardrum, get translated into electrical signals within neurons, and these signals are communicated to various other neurons in your brain via neurotransmitters. Based on these communication processes, the brain builds circuits of neurons that work together to create and store memories, initiate thought processes, produce emotional reactions, and generate actions in your body.

When scientists discovered that the messages sent between neurons were based on neurotransmitters sent from one neuron to the other, they began to develop medications that could target this process of communication. Many of the medications commonly used to treat anxiety disorders and OCD, such as Lexapro (escitalopram), Zoloft (sertraline), Effexor (venlafaxine), and Cymbalta (duloxetine), were designed to increase the amount of neurotransmitter available in the synapse, thereby affecting the circuity in the brain by enabling the neurons to change their pattern of responding. You can learn more about the way these medications affect neurons and how this influences the OCD brain in the bonus chapter "Are Medications Needed in the Obsessive Brain?" available at http://www.newharbinger.com/47186.

Modifying Circuitry Between Neurons

To rewire your OCD brain—that is, to change the patterns that the brain follows—in the hopes of having fewer obsessions, resisting compulsions, and experiencing less anxiety, we need to get technical here about your neural circuitry. Obsessions, compulsions, and anxiety are produced by the circuitry between neurons. Canadian psychologist Donald Hebb (1949) proposed a theory about how neurons create circuitry that has turned out to be very useful in explaining the process. Neuroscientist Carla Shatz distilled Hebb's theory into a simple statement: "Neurons that fire together wire together" (Doidge 2007, 63). Basically, for neurons to build connections between themselves, one neuron needs to be firing at the same time that another nearby neuron is firing. When neurons fire together, a connection between them is strengthened. If this happens repeatedly, a pattern of circuitry develops so that, eventually, activation of one neuron will cause the other neuron to activate also. More neurons can be connected to these two neurons in a similar way, and if these neurons all fire together, soon a whole set of connected neurons is created. The repetition of firing certain neurons together is what builds a circuit in the brain.

If there's one thing that's familiar to someone with OCD, it's repetition. What we don't realize is that when we repeatedly think about certain things, or carry out certain actions over and over, we are making the circuitry underlying these thoughts or actions stronger and stronger. To think a thought over and over with the hope that you'll become more able to *stop* focusing on that thought is contrary to the way the brain works. If you want to change the circuitry in your brain so that new connections form and unwanted connections are weakened, you need to understand the importance of making *new* circuitry fire. As *different* neurons in the brain fire, instead of the ones that typically fire, they are forming, and strengthening, new connections between themselves and other neurons.

Although our brains are programmed from birth to organize themselves based on connections, our brains are also amazingly flexible and exquisitely responsive to our life experiences. One person's neurons may learn to associate grandfathers with cigars, and another's neurons may associate grandfathers with farming. What you connect in your cortex depends on your experiences and what you think about. Connections between particular neurons are strengthened every time you use those particular neurons. The more frequently you activate certain neurons at the same time, the stronger the connections between them become. If whenever you see a brown smudge you think of dried blood or feces, you will make this connection stronger.

If you *don't* use certain circuitry in your brain, you'll find that the connections weaken so that you have difficulty recalling the memories you once made. This is why, if you rely on a calculator rather than your cortex to perform multiplication, you eventually may find it difficult to remember the solution to 12 x 8. And if you move to a new home and learn your new zip code, you may have difficulty remembering your old zip code. If "neurons that fire together wire together," then it's also the case that when we don't use those neural connections, we weaken them. "Use it or lose it" can apply to our neural circuitry, especially in our cortex. So if you have a tendency to focus on worries about accidentally harming a child whenever you see your nephew, the best way to change your brain to stop doing that is to focus on other things when you see him. If you talk to your nephew and learn that he likes dinosaurs, you can try to name some dinosaurs to see if he likes that kind. By focusing on dinosaurs, you're working to begin weakening the connection between your nephew and "harm" in your neural circuitry.

To stop obsessing or reduce OCD rituals and compulsions, you need to form new circuitry in your brain. We can make new connections and build new circuits in a variety of ways. When you are asked a question about your grandmother, you deliberately call to mind certain memories by activating the set of neurons holding your grandmother's image and associated memories about her. You can add new memories and images into the circuitry by being exposed to new information or images in association with your grandmother. After attending your grandmother's eightieth birthday party, you add an image of her blowing out the candles on her birthday cake to the other "grandmother" circuitry. Your brain remains flexible throughout life. New circuitry can be formed by learning to associate a definition (e.g., "placing undue emphasis on petty details") with a previously unknown word, like "pettifogging." But, if you don't rehearse the new information so that neurons representing the word and neurons holding the definition can become connected, you won't remember the definition when given the word next week. Once again, firing together promotes wiring together. Some circuitry has been rehearsed so much that it seems impossible to forget, like singing the ABCs, something even those with Alzheimer's dementia can typically accomplish.

This brings us back to obsessions. If you want to change the obsessions that you experience, you need to change the neural connections that these obsessions are based on. These connections can be in the cortex, the amygdala, or both. Connections formed by the amygdala are more resistant to change and rely on giving the amygdala specific experiences that help it learn. We'll discuss how to accomplish these neuron-based changes in chapter 9. Changing circuitry in the cortex is typically easier and can be accomplished in many ways, which we will be addressing in chapters 10 and 11. Reading, instruction, logic, experience, and rehearsal can all affect the circuitry in the cortex.

When we repeatedly think obsessive thoughts, we are making connections in our cortex stronger and more likely to pop into our minds. They are like any other thought: the more the neurons holding them are fired, the stronger their circuitry becomes. If you think about contamination every time you enter a bathroom, or think about letting people down every time you receive a request, you're strengthening these kinds of thoughts. If, whenever you see a child, you feel you must consider whether you are a danger to that child, you're making those considerations more likely to pop into your head. The circuits in the cortex operate on the principle of "survival of the busiest" (Schwartz and Begley 2003, 17), meaning that whatever circuits you use repetitively are likely to be the circuits most easily activated in the future. So, in the cortex, the more you focus on and consider certain words, thoughts, or ideas, the stronger and more frequent your focus on those particular words, thoughts, or ideas will become. Once you understand the neurology of the cortex, you realize *there is no way to decrease the likelihood of a thought by thinking it.*

If you argue with a thought in your head until you can convince yourself that you don't need to worry about it, you are just repetitively thinking about that thought. The more you focus your cortex on a particular thought, image, or urge, the stronger you're making the circuitry that holds that thought, image, or urge. Understanding that neurons that fire together wire together shows you that even arguing against your obsessions and worries just strengthens them. If you change how you think, you can change the circuitry in your brain. In chapter 10, we'll show you specific ways to change obsessive thinking by being more aware of what you think and replacing obsessive thoughts with new ones.

We've completed part 1 and covered the basics of the OCD brain, including the complex roles of the amygdala and the cortex. Now it's time to dive into the various ways you can learn to calm your amygdala in order to reduce the anxiety that fuels obsessions and compulsions. We'll begin part 2 with a discussion of the amygdala and how it creates the defense response. Making changes in the cortex sometimes results in activating the amygdala, so we'll prepare you to cope with the defense response the amygdala creates and the anxiety you experience as a result. Before we ask you to take on some of the challenges of changing your thinking processes, we want to give you some ways of coping with the stress of these challenges.

PART 2

Calming the Amygdala

CHAPTER 5

Understanding and Coping with the Defense Response

The amygdala plays a very influential role in your OCD, and you'll be better equipped to resist the demands of your OCD if you understand what the amygdala is doing. As you already learned, the amygdala is the part of your brain that produces the defense response, causing you to experience distressing emotional experiences known as fear, anxiety, or dread. These unpleasant emotional reactions provide fuel for many OCD symptoms. Anxiety increases your attention to your worries or obsessions and motivates you to engage in compulsions to try to reduce it. So by understanding what the amygdala is doing, you're seeing the root of the problem of OCD in a way you've never seen it before. The amygdala has been called the "relevance detector" (Sander, Grafman, and Zalla 2003) because it is constantly scanning whatever information your senses receive for any evidence of something that is relevant to your well-being. It monitors whatever you see, hear, feel, or smell to determine whether it is something that your attention should be directed to.

Robust evidence points to the involvement of the amygdala in OCD (Hoexter and Batistuzzo 2018). Studies have shown that when people with OCD are exposed to pictures or words intended to elicit negative emotional reactions, their amygdalae react more strongly than those of people who don't have OCD (Thorsen et al. 2018). Research has also shown a relationship between the degree of amygdala activation and OCD symptoms (Via et al. 2014).

Although we don't discuss it in detail in this book, the amygdala is involved in detecting positive things as well as negative ones. It can alert you to pleasant events and activate positive emotional responses. The warm feeling you get when you hear a song that reminds you of a loved one is because of your amygdala, for example. When you're riding the bus and feel like talking to the short, gray-haired woman who reminds you of your grandmother, the amygdala is promoting this. The amygdala plays a role in love and bonding as well as anger, aggression, and fear. But no one turns up in our offices complaining about the positive things the amygdala responds to, so we're focusing on the amygdala's role in anxiety and OCD symptoms in this book. When you feel anxiety, it's very easy to assume that there's a good reason for the anxiety, when this may not be the case at all. If you feel anxiety after touching an object, that doesn't mean you've been contaminated. Overestimating the danger associated with your anxiety and underestimating your ability to cope with anxiety fit right into the common themes that come up in people's obsessions (Davidson 2014). This chapter is an important one because here we will demystify the amygdala's purpose and influence so that you understand what your amygdala is up to and why—and are therefore less likely to misinterpret what your symptoms mean.

We explained in chapter 2 that the way the brain is organized allows the amygdala to be both an early warning system and a defensive coordinator, able to signal danger and set in motion protective responses before the cortex even has time to process the information coming into the senses. Knowing the specific responses the amygdala sets in motion can help you interpret your anxiety correctly and cope with it better, making you feel more willing to resist your obsessions and less likely to get caught up in compulsive behaviors like checking or seeking reassurance.

Remember that the amygdala is a defense coordinator, working to protect you from potential threats, so the responses that it creates are all intended to protect you. As the amygdala is receiving information from your senses, if it detects a threat of any kind, it sends messages to other parts of your brain to produce that whole set of changes in your body that we referred to earlier as the fight, flight, or freeze response. Although Walter Cannon (1929) first recognized the fight-or-flight pattern of responding, it was endocrinologist Hans Selye, in the 1930s, who investigated what happens in the body when this pattern of responding occurs. By studying rats, Selye examined the way the body reacts when animals (including humans) are under stress and discovered that the reaction in the body follows a very specific, recognizable pattern in all animals (Sapolsky 1998). Selye called this response the "stress response," but we think the term "defense response" is more accurate because it reminds us of the amygdala's purpose.

When it detects a threat, the amygdala can rapidly send messages to some very influential structures in the brain, including the brainstem and the hypothalamus (Asan, Steinke, and Lesch 2013). By communicating with these structures, the amygdala can instantly cause a series of changes in your body, energizing the sympathetic nervous system, activating movement systems, increasing levels of neurotransmitters, and prompting the release of hormones like adrenaline and cortisol. Remember, however, that the amygdala can respond in error—all of these responses can be triggered unnecessarily. One of the most important bits of knowledge you should have learned about the amygdala is that it can be *wrong*. This is hard to remember because, when the defense response is triggered, it *feels* like you are in danger. This is the essence of anxiety. The feeling is very real— don't let anyone tell you otherwise—and we can even identify bodily changes that prove it's occurring. But *the feeling does not mean that there is a danger.* Your amygdala and your body can be creating physiological responses that lead to a feeling of danger when no true danger exists. This feeling of danger can make you get caught up in obsessive thinking or rituals that aren't needed—because no danger exists.

Experiencing this feeling of danger or dread is a central problem for those with OCD. Note that here we are focusing on the *feeling* that you're in danger, not the *thought* that you're in danger. These are two separate things. Take a very simple example: Imagine you're in the grocery store choosing a bunch of bananas when someone unexpectedly shouts, "Hey!"

behind you, and you nearly jump out of your skin. That is a *feeling* produced by your amygdala. Some thoughts may also come to mind, like *Should I not have touched those bananas?* or *I might be in some danger!* but those thoughts are separate from the *feeling* of being startled or frightened that you experience in your gut or chest. As you turn around, eyes wide, you see a friend you haven't seen in months, who gives you a hug, saying, "I'm so happy to run into you! How have you been?" Obviously, the feeling produced by the amygdala was in error and was not needed in the situation. But the feeling was very real, and we often trust that feeling to be accurate much more than we should.

This is a very difficult insight to grasp, especially for those with OCD, because of two tendencies of the OCD brain: to overestimate threats and to give too much importance to thoughts and feelings. But the amygdala's reactions should not be trusted for two important reasons: (1) they can be based on incomplete information and (2) they are not always useful. Compared to the cortex, the amygdala has limited information. Further, most of the bodily reactions have to do with preparing the body to flee, fight, or freeze, strategies that may have been crucial for our ancestors but are not always helpful in our twenty-first-century lives. While we often cannot stop these bodily reactions from occurring because we can't control the amygdala, we don't have to *trust* that we need to run away or protect ourselves just because of that impulse. We don't have to do anything (like checking or counting) in order to keep ourselves safe; we can just wait for the feeling of anxiety to end. That may seem to be a very strange approach, but it works!

Take some time to review figure 6. You can see a whole set of physiological responses that are part of the defense response that the amygdala can put into action (LeDoux 2015). Underlying these responses is the amygdala's FFF response; its goal is to prepare us to fight or flee or freeze. Increased blood pressure and blood supply from the heart, increased muscle tension, and the release of glucose to fuel responses in the muscles occur, enabling you to use your arms or legs. At the same time, digestive processes are slowed because blood is being rerouted to muscles in the extremities, so you may feel a queasy feeling in your stomach as this occurs.

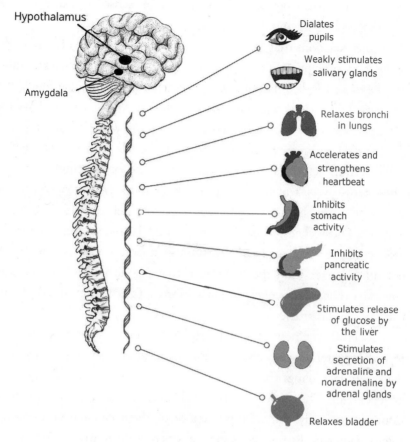

Hypothalamus

Amygdala

Dialates pupils

Weakly stimulates salivary glands

Relaxes bronchi in lungs

Accelerates and strengthens heartbeat

Inhibits stomach activity

Inhibits pancreatic activity

Stimulates release of glucose by the liver

Stimulates secretion of adrenaline and noradrenaline by adrenal glands

Relaxes bladder

Figure 6

Some reactions that are activated by the amygdala, including dilation of the pupils or the release of adrenaline, are ones most people recognize as helpful if they are in danger. Better vision in the dark or a burst of energy could be helpful in a dangerous situation. But other very disconcerting reactions—like feeling the need to urinate or, even worse, diarrhea, can also be connected to the defense response. Yes, the amygdala can cause you to make a quick run to the bathroom when you're under stress. We see this in our pets too! When Catherine takes her dog to the vet, she gives him time to relieve himself before entering the clinic. Otherwise, poor Buddy makes a mess as the vet techs pull him down the hallway to trim his nails. Our clients often laugh when we point out that relieving ourselves when we're in danger makes sense: we can run faster when we aren't carrying that extra weight!

Other reactions that go along with anxiety—like a lack of stimulation of the salivary glands, which causes a dry mouth, or loss of sexual arousal—may be more perplexing. But it seems pretty clear that people should not be focused on eating or sexual activities when they are in danger! We are the descendants of the frightened people—those who had an amygdala that reacted with fear to certain situations. People with amygdalae that made certain activities (like eating, having sex, or sleeping) less likely when danger was detected were probably more likely to survive because they kept aware of dangers and lived to pass on their genes to future generations. Those ancestors gave you the amygdala that you have, faithfully producing the good old-fashioned defense response whenever it detects potential danger. Add the knowledge that people with OCD have an amygdala that is *more* likely to react strongly (Thorsen et al. 2018), and you can see the problem.

Hopefully you now understand that the symptoms of anxiety you feel in your body reflect the amygdala's misguided attempts to rescue you from danger. In fact, danger often does not exist, even though bodily reactions are occurring. The tendency of the OCD brain to take feelings and sensations very seriously, and to focus attention on them, can make these symptoms a source of distress. For example, when Jason is driving and hears a horn honk near him, his amygdala produces the defense response, and Jason experiences a release of adrenaline, a pounding heart, and a sense of danger. If instead of interpreting these symptoms as a protective response from his amygdala, Jason overestimates their threat and begins to worry he may have a heart attack or lose control of himself, he may think he cannot drive safely and believe he must get off the road. The problem is not that Jason's amygdala reacted. The problem is Jason's interpreting his physical sensations and thoughts as more dangerous and important than they are.

When you have OCD, you are very vulnerable to developing obsessive thoughts and worries focused on these bodily reactions. When you feel a change in your body, do you focus on it and try to interpret its meaning? It's perfectly normal to do so, especially since the amygdala produces these reactions without your having any awareness of their creation or having any control over them. You may wonder why you are trembling or what that nauseous feeling in your stomach may be due to. Also, you can't necessarily stop these physical reactions, at least not quickly, and that lack of control is troubling. If you aren't aware that these responses are normal aspects of the defense response, then it's easy to feel that some of these physical symptoms should be a source of concern.

Experiencing the Defense Response in the Body

When we experience fear or anxiety, what are we experiencing in our bodies? Recently a friend was facing a lawsuit and was embarrassed that he couldn't keep his hands from trembling as he sat in the courtroom. He worried that this trembling might mean that he was guilty, even though he knew he was being falsely accused. You can see how the trembling was fueling his obsessive thinking about being guilty, even though these bodily reactions are a normal part of the FFF response. The muscle movements reflect the amygdala's activation of areas in the motor system that can produce running or jumping behaviors, and my friend was relieved to know why they were occurring. (He was also grateful to learn some ways he could work to calm the amygdala in order to stop them.)

Unfortunately, when these responses are triggered by the amygdala, people with OCD are very prone to misinterpret them. This is particularly true of responses related to increased heart rate, stronger heart contractions, or alterations in breathing. Many people interpret these reactions to suggest that they're experiencing a threatening health event like a heart attack or stroke. They may obsess about something being *wrong* with their body, when in actuality, the changes in the body reflect the functioning of a very strong, healthy body that's ramping up the defense response. Because Doug's heart is pounding as he reviews his finances, wondering how to pay his mortgage, he worries he's having a heart attack. But, as we often remind our clients, when your heart is pounding strongly, that is a *good* sign that your heart is healthy, not a sign that your heart is about to stop. You have a healthy body, not an illness! Misinterpreting the amygdala can lead to obsessions about physical health. A basic understanding of the FFF response is very useful in reducing some of the obsessions or worries that develop in OCD. Even though every once in a while fighting or fleeing do help, usually they don't fit the situation we're in, and we overthink our bodily reactions to try to make sense of them. We need to remind ourselves that the amygdala is creating these reactions and not misinterpret them or seek reassurance about them.

Remember that the amygdala can initiate a whole set of bodily responses very rapidly, before you have even completely perceived what the amygdala is responding to. The amygdala acts so swiftly that you *cannot* control it, nor can you control the bodily reactions it produces. People who try to control and manage them often will feel that they're somehow failing at being in control of themselves. Although some coping strategies can help return the body to a more relaxed state, as we will explain in chapter 6, the truth is that we are limited in our ability to control the amygdala and its effects. Many times, symptoms of anxiety are going to occur, whether we like it or not. Mindfulness and acceptance are strategies that can help you shift from trying to be in control to simply focusing on the physical experience with curiosity and acceptance while you let it pass. We'll discuss these strategies in chapter 10.

Many aspects of the defense response are misinterpreted. When people tremble or find themselves frozen in fear, they may think they're losing control of themselves or going crazy. If you have OCD, you may overthink the experience and may obsess that you could lose control and carry out an act that you don't want to commit. After all, you're behaving in a way that you have no control over. What else are you capable of doing? Because behaviors associated with the FFF response don't make sense to us, the cortex may think about it (or ourselves) in ways that produce even more anxiety. The cortex likes to try to make sense of what is happening, but in coming up with explanations, it may make anxiety worse. As Maria prepares to meet with her teacher about how to improve her grades, her legs start bouncing and she can't keep them still. She makes the situation worse by thinking about how the teacher will interpret her hyperactive legs and starts thinking that she can't talk to her teacher as long as she's feeling this way. This is why a basic understanding of the FFF response is so useful in reducing some of the obsessions or worries that develop in response to it. When you know that the amygdala is creating these responses in an attempt to protect you from danger as part of a normal, healthy FFF response, you recognize the meaning of your symptoms and don't come up with incorrect interpretations. If Maria goes for a brisk walk, it will help burn off some of the adrenaline fueling the FFF response, and if she keeps her mind on the idea that the teacher offered to help her, rather than on her legs, she'll be more likely to get a better grade, which is what is important to her.

Some of the responses the amygdala produces, such as the release of adrenaline, have effects that can last for a period of time, even after the amygdala stops reacting to a potential threat. You can't take back the adrenaline that has been released, and for five or more minutes you may still feel its effects. But a brief burst of adrenaline is perfectly normal and is not dangerous. You may have a tendency to focus on the feeling and what it means, and relate it to your obsessions, in which case it fuels your obsessive thinking. It may help to ask other people whether they experience these reactions. You may be surprised to find out that you have friends who tremble or feel the effects of adrenaline for several minutes after they are startled. The difference between you and your friends may be that you focus on these reactions much more than they do and relate them to your concerns, while they just ignore the sensations and move on.

Remember that when the amygdala takes charge, especially if you start to feel particularly panicky, it is able to override other parts of your brain, making it difficult for you to focus your attention or think clearly. As we discussed in chapter 2, it's normal to have difficulty focusing or thinking when the amygdala is activated. Joseph LeDoux (2002, 226) referred to the amygdala's ability to dominate brain functions as a "hostile takeover of consciousness by emotion." This means that, as you calmly read this book, you'll be more able to understand what the amygdala is causing to occur in your body. When you're experiencing anxiety, however, your ability to focus your thoughts will be more limited. The higher the level of anxiety you experience, the more difficult it will be to focus on interpreting your symptoms with a calm intellect. So when you're experiencing the bodily reactions of the defense response and jumping to frightening conclusions that there is a true danger, try to use coping thoughts that are fairly simple and straightforward. Your cortex may be off-line and not able to process complex ideas about fight or flight, adrenaline, or shifting blood flow. Reminding yourself, *This is the normal defense response; My amygdala is creating a normal, healthy reaction*; or *This response is not dangerous* is a good approach when you see yourself trembling or feel your heart pounding. Whether your fear is that you have been contaminated or that the piles of papers on your desk mean you are incompetent, accept that your bodily reaction does not mean that your fear is justified. You may not be able to focus on complex tasks because your amygdala is activated, so find something productive you can do that is relatively simple. When Jessie is overwhelmed with anxiety by the piles of papers on her desk, she focuses on doing simple, structured tasks like paying bills online or opening the mail and sorting it into piles of junk mail to be recycled or bills to deal with later. She knows this is not the time to call a customer to discuss a complicated issue. She accepts that the amygdala has the upper hand for the moment but continues to focus on her job in whatever way she can.

How Do You Experience the Defense Response?

Take some time to consider how you personally feel the defense response, or observe yourself in different amygdala-activating situations to see what kind of bodily reactions you experience when it occurs. The defense response can be fairly mild at times and just be reflected in muscle tension or a dry mouth, or it can be as extreme as a panic attack, when your heart pounds very strongly and your body is literally shaking. All of the symptoms below are normal for people to experience when the amygdala activates the defense response. Many of them reflect activation of the sympathetic nervous system. Read through the list and check those that apply to you.

_____ Pounding heart

_____ Changes in breathing (rapid or holding breath)

_____ Need to urinate

_____ Diarrhea

_____ Desire to flee or withdraw

_____ Stomach distress

_____ Dry mouth

_____ Flushing

_____ Light-headedness

_____ Muscle tension

_____ Immobilization

_____ Trembling

_____ Perspiration

_____ Difficulty focusing

Recognizing any of the symptoms above in your own reactions indicates that your amygdala is doing its job to try to protect you. The FFF response is essential in preparing you to respond immediately to emergency situations. It may have saved you from a car accident or helped you defend yourself from some other danger. These reactions are hardwired into you, and they will occur naturally. The defense response occurs even when it doesn't help—when you're worrying you can't afford the cost of a recommended surgery or when your teenage daughter throws the blow dryer on the floor and breaks it. These symptoms are the amygdala's way of protecting you, and if you want to have some control over them, learning about how the amygdala works is your best defense.

A panic attack—one of the most unpleasant experiences a person can have—has its roots in the amygdala. _Panic attack_ is the name given to the defense response when it seems much stronger than needed or inappropriate to the situation. (If you have a very strong defense response in a situation in which you need it, such as when you're attacked by someone, we call that a useful defense response!) If you have an extreme defense response because you feel something slimy on your hand as you prepare supper, we would call that a panic attack. Symptoms of a panic attack can include feelings of panic or terror, a pounding or racing heart, hyperventilation, light-headedness, an urge to flee, trembling and shaking, or numbness and tingling. People can also have nausea, sweating, an urgent need to go to the bathroom, hot flashes or chills, feelings of paralysis, shortness of breath, and discomfort in the chest. Thoughts can be affected also; some people think they are going crazy and some have a sense of unreality.

You may recognize these symptoms as similar aspects of the defense response created by the amygdala; they are simply a very extreme defense response. When people experience a panic attack, some may feel like attacking others, some may feel like running away, and some may be paralyzed—consistent with the fight, flight, or freeze response. Most people will have at least one panic attack in their life. Some people seem to be born with an amygdala that makes them prone to developing panic attacks, and some people develop panic attacks because of the response of their amygdala to traumatic experiences.

Once a full-blown panic attack starts, you can do very little to stop it. The physiological processes that have been initiated must run their course, and your cortex can't stop them. If you're having a panic attack and someone tries to reason with you, explaining logically that you aren't in danger, he is essentially talking to a cortex that is turned off. You are unlikely to be able to process very complex information; the best you can do is make a panic attack shorter. On the other hand, shutting down a less extreme defense response is more possible than shutting down a panic attack, and we will identify some strategies to do so in the next section.

Strategies to Cope with the Defense Response

Throughout this chapter, we have been providing you with knowledge about the defense response to make sure that you don't misinterpret defense-response-related symptoms and obsess about them in ways that worsen your OCD. Making sure you correctly recognize and interpret the signs of the defense response is a task for your cortex. It's bad enough coping with the defense response; we don't need the cortex misinterpreting what's happening in our body as a sign of danger and thereby activating the amygdala even more. We learned in chapter 3 that the cortex does not have the connections to create the defense response directly—only the amygdala can do that—but the cortex can activate the amygdala to *worsen* the defense response by creating frightening thoughts and images. In this section, we'll focus on ways for you to use your cortex and your amygdala to cope with your defense response. All of these methods can help reduce anxiety, taking away the fuel that drives your obsessions and compulsions and putting you more in control of your life.

We'll begin with changes you can make in your cortex. As you experience a defense response, some disturbing thoughts and worries can come into your cortex; frequent thoughts are about becoming ill or causing harm to someone. These kinds of thoughts not only can become worries or obsessions, but they can also worsen the defense response or make it last longer. Here are some steps you can take to avoid this.

Take care not to misinterpret the bodily reactions of the defense response. Be able to identify the common symptoms you get when you experience the defense response or a panic attack so that you can remind yourself that they are normal bodily responses produced by the amygdala when it (rightly or wrongly) identifies a threat. In particular, do not misinterpret symptoms as indications of a catastrophic or serious illness or as an indication that you're going to lose control of yourself and carry out some violent or dangerous act. If you focus on these kinds of thoughts, you're likely to worsen the defense response and your feeling of anxiety. "What if" questions can often lead you down a path of obsessing about the meaning of certain symptoms. If you've had the symptoms before, remind yourself that, when they occurred, they did not mean that you had an illness or would lose control.

Remember that feelings of dread or anxiety do not mean there is a true danger. Because the defense response is typically accompanied by a feeling of dread or anxiety, and a panic attack is accompanied by feelings of panic and fear, it's easy to take these feelings to mean that something is terribly wrong or truly dangerous. People with OCD are prone to overestimate the importance of thoughts or feelings, but a feeling can be very real and very wrong (about danger) at the same time. You can have terrible anxiety before you ask someone to marry you, but, despite the anxiety, find that you are soon engaged to an ecstatic partner. Feelings do not predict what will happen, and this is especially true of anxiety. Just because your amygdala is reacting as if there is a danger does not mean a danger truly exists.

Don't watch for, or anticipate, anxiety or panic attacks. If you're constantly expecting and worrying that you'll become anxious, frequently checking for symptoms, or being vigilant for any unusual sensations, you're more likely to create these symptoms. Your vigilance will cause the amygdala to begin expecting danger and will activate the defense response. "What if I panic?" questions are likely to lead to thoughts and images that activate the amygdala and actually make a panic attack more likely to occur. Keep your focus on what you need to do and on your surroundings in the moment rather than thinking too far ahead or worrying about what could occur.

Practice acceptance rather than attempting to control symptoms. Many people with OCD have a strong desire to be in control of all kinds of things—so it's easy to focus on trying to control amygdala-related symptoms. But when you focus on being in control of the defense response, you're setting yourself up for failure. Physiological processes outside your control create your defense response. By understanding why you have these physical responses and what they mean and don't mean, you reduce the uncertainty about them. But if you get into a battle trying to control them, you introduce uncertainty again. *How long will this last?* The symptoms tend to go away sooner when you accept them. Simply notice the experience and acknowledge it: *My heart is pounding.* It doesn't need to go away for you to focus on other things. And you don't need to pretend you're happy about a symptom to accept it. *I hate my amygdala making my legs bounce like this. My amygdala is just making a big deal about nothing!*

Distract yourself. When you distract yourself, you're replacing one set of thoughts with something else. As we noted in chapter 3, if we want to stop thinking certain thoughts, we need to replace them. What to think about? Anything! Think about what you need to accomplish that day, or pay attention to the television or a podcast. Call a friend. Make a grocery list. Imagine what kind of vacation you'll take. Do your job. Wash the dishes. See how many different green objects you can find in your surroundings. Appreciate the fall colors. Just live your life. Select something you can get done in the next fifteen minutes and focus on that. There is no good reason to keep thinking about or worrying about the nauseous feeling in your stomach or other symptoms. *It's bad enough the amygdala causes me to experience these symptoms—I don't want to give them more attention than they deserve.*

We should note that these cortex-based coping strategies are most helpful when what you're experiencing is the defense response. But during a full-blown panic attack, cortex-based coping strategies are often not very useful. In a true panic attack, you're likely to be too anxious to think clearly as the amygdala hijacks your cortex. During panic attacks, the best solution is to breathe deeply and slowly, try to relax your muscles, and distract yourself as best you can as you wait for the panic attack to run its course. Exercising also can help with a panic attack. Panic attacks always pass, and using breathing and relaxation strategies can hasten their passing. What extends them are thoughts that activate the amygdala further: worries about heart attacks or losing control, for example. Also, during a panic attack, you don't have access to your cortex in the usual way, so you shouldn't engage in activities that require focusing on your thoughts or carefully controlling complicated behaviors. You shouldn't attempt to make critical decisions or engage in complicated behaviors like driving during a panic attack.

We now turn to coping methods that help us avoid or shorten the defense response by focusing on changing the amygdala. We've been saying that the amygdala is difficult for us to directly control. In the same way, many of the physical responses that the amygdala puts into motion during the defense response are difficult, if not impossible, to control. Once adrenaline is released into our bloodstream, we cannot stop its effects, for example. But there are some coping strategies that do help, and we identify them here. (In later chapters we'll elaborate on these techniques.) Each of these coping strategies has been shown to reduce amygdala activation in a matter of minutes. Because the amygdala has such an important influence in OCD, anyone with OCD can benefit from knowing ways to reduce amygdala activation.

Deep breathing: A specific type of breathing has been shown to have effects that *directly calm the amygdala* (Taylor et al. 2011; Zelano et al. 2016): slow, full, deep breaths. Using deep breathing reduces amygdala activation, so it helps you have some control over the amygdala in a way that few other interventions can. In addition, it's cheaper than medication and can be used almost anytime and anywhere.

Muscle relaxation: When the amygdala produces the defense response, muscle tension increases. Tight muscles seem to further increase amygdala activation, whereas relaxed muscles are more likely to calm the body down. Learning to recognize when your muscles are tense and diligently practicing specific techniques that relax your muscles will help reduce the defense response. Muscle relaxation, especially when combined with deep breathing, has been shown to effectively reduce anxiety (Neeru et al. 2015).

Exercise: When you're anxious, stressed, or panicky, exercise can reduce anxiety in only twenty minutes (Johnsgard 2004) and counteract the defense response and panic attacks by burning off the excess adrenaline in your system. If you remember that one purpose of the defense response is to get the body ready to fight or flee, you'll recognize that physical exertion is exactly what the body is ready for. Aerobic exercise that involves large muscle groups seems to be most helpful in reducing amygdala activation (DeBoer et al. 2012). Even a brisk twenty-minute walk can calm the amygdala down. Exercise works even when nothing has occurred to change whatever provoked the amygdala. You can still have a distressing mess in the kitchen, but you just feel calmer about it.

One final point about the defense response is especially important to those with OCD. When you feel dread, anxiety, or especially panic, you'll likely turn to some compulsive behavior to cope with the situation. When you do this, however, you're strengthening OCD's control over you. By increasing your understanding that your anxiety is really a result of the defense response, we hope to help you resist getting stuck in obsessions and also reduce the pull of your compulsions. Anxiety is the fuel for compulsions because we engage in compulsions to reduce anxiety. If you can learn other methods to reduce anxiety, and learn to accept and tolerate it when you can't reduce it, you'll find that it always goes away eventually.

Even though compulsions—such as washing your hands, seeking reassurance, or going back home to check—may help reduce your anxiety in the short term, in the long term these compulsions are becoming stronger each time you give in to them. Compulsions are not a productive solution because they tend to rob you of both time and control in your life. They also sustain the amygdala's tendency to produce the defense response and cause you more anxiety because the amygdala doesn't have the opportunity to learn that no danger exists.

In this chapter, you've learned about the amygdala's important role as a defense coordinator and its effects on your body. You've also learned strategies to cope with the defense response produced by the amygdala. Whenever possible, use the cortex-based coping strategies we discussed above while you relax, breathe, and accept the defense response without giving in to your compulsions. When the defense response produces feelings of anxiety, don't let that anxiety lead you to believe your compulsions are necessary. As you have learned, your amygdala may be responding as if you are in danger, but there is likely no true danger at all.

CHAPTER 6

Calming the Amygdala with Relaxation Strategies

When you have OCD, uncertainty is so uncomfortable, and the tendency to overestimate threats makes you want to manage everything...so you might have a strong need to be in control. You've learned that the amygdala does many things that are beyond your control, and we're encouraging you to learn to accept that. Of course, this is a challenge. But our goal for you is not the impossible task of controlling the amygdala, but instead, taking control of your life back from OCD. That goal is much more achievable, especially if you keep in mind the Serenity Prayer: *Grant me the serenity to accept the things I cannot change, courage to change the things I can, and wisdom to know the difference.* We will help you learn what you can and cannot change in the process of taking control back from your OCD.

To be in charge of your life, rather than allowing your OCD to control you, you'll need to learn some strategies to manage your amygdala. Remember, the amygdala is the root of the anxiety that leads you into obsessive thoughts that dominate your thinking and the compulsions that take over your life. Using compulsions to cope with your anxiety is only a temporary fix for your anxiety; you feel better after engaging in a compulsion, but only temporarily. Then you have to do it again and again. To take back control of your life, focus on the amygdala, the source of anxiety. In the last chapter you learned strategies to cope with the defense response when the amygdala becomes activated. In this chapter, you'll learn effective strategies to reduce the number of times the amygdala is activated—and to therefore experience less anxiety in the first place. This allows you to take control of your life back from OCD.

Consider Manuel, who has to give a presentation to a large group of executives. As he puts together his PowerPoint presentation, he obsesses over his margins and checks and rechecks his grammar. Manuel needs to do things correctly and finds himself dissatisfied with every sentence he puts on his PowerPoint. He starts to doubt his ability to do this presentation well. He notices his heart is pounding and he feels warm and sweaty. He begins to think that if he's sweating like this just putting together the PowerPoint, he's not going to be able to present his ideas well. He doesn't want to be sweating, flushing red, or trembling in front of these executives. Manuel should not be wasting precious time obsessing over everything that could go wrong. His physical reactions are coming from the amygdala's defense response activating the sympathetic nervous system (SNS)—the system producing the FFF response—and causing a release of adrenaline as he imagines himself having difficulties giving his presentation. Can Manuel do anything about these reactions?

Fortunately, a variety of techniques can be used to activate the parasympathetic nervous system (PNS), which reverses many effects of the SNS. Whereas activation of the SNS creates the FFF response, the PNS's action is often referred to as "rest and digest." It slows the heart rate, reduces blood pressure and blood flow to the extremities, and increases digestion through the secretion of saliva, mucus, and bile. Using techniques to increase parasympathetic responses can decrease sympathetic responding.

The PNS is more likely to be activated when people are relaxed. That's why therapists often encourage anxious clients to use *relaxation strategies*, one of the primary methods used to facilitate PNS activation and decrease SNS activation. The goal is to reduce the processes of the defense response that the amygdala has put into motion. Relaxation has been used for years, ever since psychiatrist Edmund Jacobson (1938) developed a process called progressive muscle relaxation, which his patients found to be very beneficial at combatting the defense response. Once imaging techniques like fMRI and PET scans were developed for directly observing processes in the human brain, researchers discovered that when people practice relaxation strategies, activation in the amygdala itself is reduced! A variety of studies have shown that techniques that promote relaxation, such as breathing exercises and meditation, reduce activation in the amygdala (Jerath et al. 2012; Taylor et al. 2011). In other words, when you increase PNS activation, you not only reduce the defense response, but you also reduce activation of the amygdala, so you are intervening at the source of the problem.

Teaching yourself to relax is one of the quickest, most effective ways to calm the amygdala. Neuroimaging studies have shown that changes in the brain occur when people practice various relaxation strategies, like breathing exercises (Goldin and Gross 2010), meditation (Desbordes et al. 2012), yoga (Froeliger et al. 2012), and even chanting (Kalyani et al. 2011). Many of these approaches reduce activation in the amygdala in minutes— even faster than the prescribed medications that reduce activation in the amygdala most quickly, the benzodiazepines (e.g., alprazolam, diazepam, clonazepam), which can take thirty minutes to have an effect. Moreover, these approaches do not have the negative side effects that the benzodiazepines have. (For more information about benzodiazepines, see the bonus chapter "Are Medications Needed in the Obsessive Brain?" on the website for this book at http://www.newharbinger.com/47186.)

Returning to Manuel, we can offer him some strategies to calm his amygdala and reduce his defense response. We begin by explaining to Manuel that his physical symptoms reflect the operation of a healthy defense response elicited by his imagining himself giving a presentation, and they are not a prediction of how his presentation to the executives will go. If he uses a relaxation strategy, like deep breathing, and keeps his attention on putting together a compelling presentation (rather than obsessing about what could go wrong during the presentation), he'll reduce activation in his amygdala, decrease his amygdala-activated symptoms, and build his confidence in the presentation. He needs to accept that he'll have some symptoms of the defense response, which are completely normal, and focus on the content of his presentation without getting pulled into perfectionistic obsessions about needing control over bodily reactions that may not even occur when he actually gives his presentation.

We'll present several forms of relaxation training in this chapter. Every person is different, and different strategies can be more appealing or more effective for one person than another. No matter which of these approaches you choose, each has scientific evidence showing that the strategy will reduce the defense response and anxiety created by your amygdala. We need to accept that, because of the way our brain is wired, the cortex has very little direct control over the amygdala. That means that we do not have *conscious* control over the amygdala's functioning. Telling ourselves, or the amygdala, to calm down will not have much effect. These relaxation strategies give us much needed ways to influence processes that the amygdala activates, and, in some cases, they directly impact the amygdala itself.

Even though individuals respond differently to various relaxation strategies, virtually everyone will benefit from relaxation training if they practice. Most approaches to relaxation focus on two physical processes: breathing and muscle relaxation. When we therapists explain that breathing can help reduce anxiety, the defense response, and even the effects of a panic attack, our clients are typically skeptical. How could something as simple as breathing differently give a person influence over these processes? Luckily, we have fMRI studies that show the brain in action and allow us to see how different parts of the brain are activated in real time. These studies show a reduction in the activation of the amygdala occurring within minutes after a person has begun deep breathing strategies (Doll et al. 2016; Goldin and Gross 2010; Taylor et al. 2011; Zelano et al. 2016).

Once you try these specific breathing strategies, their usefulness is often immediately apparent. When Lexi came in for a session one day and her therapist encouraged her to try some relaxation breathing, she was skeptical because it sounded so irrelevant to her concerns. She was stressing about her constant self-questioning, and what she really wanted was reassurance about her dissertation, not a breathing exercise. But after breathing as directed for several minutes, she reported that her anxiety had definitely decreased from the level 8 she walked in the door with to a level 3. "I didn't know just breathing could do that!" she exclaimed. She was willing to turn away from her tendencies to obsess and seek reassurance in order to find a more effective way to deal with her OCD.

Breathing-Focused Relaxation Strategies

The type of breathing shown to be most effective in *directly calming the amygdala* in a way that few other interventions can is slow, deep breathing. So it really is worthwhile to learn how to use this technique if you want to exert some control over the amygdala.

Try this breathing exercise:

1. Take a moment to focus on your natural breathing right now.
   - Are you holding your breath?
   - Are you breathing deeply or shallowly?

2. Take a deep breath, making sure to let in as much air as possible.
   - Try to completely expand your lungs, including the lowest parts.
   - Don't gulp air. Inhale slowly and deeply.

3. Breathe all of the air in your lungs out.
   - Breathe out slowly; pursed lips can help keep it slow.
   - Make sure your lungs are completely empty before the next breath.

4. Repeat this inhale/exhale routine several times.
   - Don't hold your breath; at any moment you are either breathing in or breathing out.

5. Now focus on slowing the speed at which you breathe using a clock or watch.
   - Breathe *slowly* in and out to time your breaths to five or six breaths a minute.
   - One breath = one inhale and one exhale.
   - Use this breathing rate for five to ten minutes to promote PNS activation.

What did you learn from this exercise? Did you find that this type of breathing had any emotional effect? Did you catch yourself holding your breath at times? Did you focus more on inhaling or exhaling?

Some people immediately feel a reduction of anxiety and stress when they do this type of breathing for a period of time. Adopting a slow rhythm of deep breathing for at least ten minutes can be soothing and relaxing. When experiencing something stressful, people tend to hold their breath or to breathe shallowly, without any awareness that they are doing this. This type of ineffective breathing is very likely to activate the amygdala. Deep breathing corrects this. Breathing out is especially important; strong exhalation is more likely to activate the PNS.

Breathing techniques are so simple and inexpensive that many people underestimate their effectiveness in influencing the amygdala and the effects they create in the body. Deep, slow breathing has a strong impact on the PNS. The pace at which you breathe is very important. If you breathe slowly and deeply at the rate of five or six breaths per minute for even five minutes, you can often feel an improvement in your stress and tension level. A fifteen-minute session can provide a reset to the whole defense response.

The amygdala is very sensitive to the level of carbon dioxide (CO_2) in our bloodstream, which goes up when our oxygen supply is depleted, when we aren't breathing effectively, or when we are exposed to CO_2 in our environment. Researchers have called the amygdala a *chemosensor* (Ziemann et al. 2009) and have shown that the amygdala monitors CO_2 levels and will activate very quickly if they are high. Therefore, breathing out CO_2 and getting fresh oxygen into your body is an important tool in keeping the amygdala calm. Important note: If you ever want the amygdala to create panic, just find a way to cut off your air supply, and you can guarantee you will activate your amygdala!

When you use deep breathing, not only are you calming the amygdala and reducing the defense response (decreasing blood pressure, lowering heart rate), but you are also improving mental clarity, releasing toxins, easing pain, and releasing endorphins that elevate your mood. This is something you can do wherever you are and while doing almost anything, and others don't typically observe it. You don't have to breathe with your mouth open; it doesn't matter if you breathe through your nose or mouth.

The best way to use relaxation breathing is to incorporate it routinely into your daily life. You should do at least ten to fifteen minutes of deliberate deep breathing at least a couple of times a day, and add some additional deep breathing when you recognize you need to calm your amygdala in a specific situation. You can think of the process as similar to having air conditioning in your home—air conditioning needs to come on repeatedly during the day to keep your home cool. You need to keep resetting your amygdala down to a lower "temperature" multiple times during the day, with targeted relaxation breathing for stress-provoking situations. If you feel anxiety as a result of intrusive thoughts, worries that you have been exposed to contaminants, or doubts about whether you offended someone, relaxation breathing can help you keep your amygdala under control. You reduce the fuel for those obsessions. If you have compulsions—to check, wash your hands, or get reassurance—relaxation breathing helps you resist these compulsions by reducing the defense response and thereby reducing the anxiety that fuels the compulsions.

Not everyone finds deep breathing helpful. If a person has asthma or other breathing difficulties, focused attention to breathing can sometimes increase anxiety. If this is true for you, you may find greater benefits from the relaxation strategies that focus on reducing muscle tension. If you avoid using breathing strategies, just make sure that, at a minimum, you notice whenever you hold your breath, and make sure you're getting enough oxygen. You may be surprised how much holding your breath (which seems related to the freeze response) causes amygdala activation. Breathing is essential for a calm amygdala!

Avoid Hyperventilation

Hyperventilation is overbreathing that can result in a variety of symptoms. If you breathe deeply at a rate of five or six breaths per minute, hyperventilation is unlikely to occur. But it is good to know the signs of hyperventilation. Sometimes a person who overbreathes will feel faint or dizzy. Hyperventilation can also result in a drop in calcium levels in your blood; numbness or tingling in your face, arms, and hands; or muscle twitching and cramping in your hands and feet. You can also swallow excessive air, resulting in belching, pressure in your stomach, or a dry mouth. If any of these symptoms occurs, you may mistakenly consider them to be a part of the defense response or due to anxiety, but they can be corrected by ensuring that you're breathing in a healthy manner. Often we can correct problems with our breathing by simply being more aware of the pace of our breath and making sure that we neither hold our breath nor hyperventilate.

People who hyperventilate were once advised to breathe into a paper bag, but that approach has been discarded. It seems to be more effective to simply breathe deeply and slowly. Rebreathing the same air that you're breathing into the bag increases the level of CO_2 that you are breathing, and that can lead to problems.

Diaphragmatic Breathing

We all know how to breathe, and we spend little time thinking about it, but different types of breathing can be trained. One type is recommended for its particular effectiveness in activating the PNS to counter the amygdala's activation of the SNS. *Diaphragmatic breathing*, or abdominal breathing, is when you breathe more from the abdomen than from the chest, and the movement of the diaphragm (the muscle under the lowest part of the lungs) is put into motion.

To practice diaphragmatic breathing, sit comfortably in a chair with your feet on the floor. Place one hand on your chest over your heart and the other hand on your stomach above your belly button. Breathe normally, and see which part of your body expands as you inhale and exhale. Simply looking down at your hands as you breathe can help you see this. If you're breathing shallowly, you're likely to feel and see movement in your chest more than your abdomen. Now focus on breathing deeply in a manner that can completely fill your lungs. This should cause expansion in your abdomen. As you breathe deeply, look down at the hand on your abdomen and consciously try to breathe in a way that expands your abdomen so that you can see the outward movement. Many people unknowingly pull their stomachs in as they inhale, and this keeps the diaphragm from expanding downward in order to allow the lungs to completely fill with air.

With regular practice, you can modify your breathing to be slower and deeper, and to be more diaphragmatic and effective. Practicing diaphragmatic breathing for five to ten minutes just three times a day can improve your awareness of healthy versus unhealthy breathing. During your day, also watch for times when you notice you're holding your breath, breathing shallowly, or hyperventilating, and practice establishing a better breathing pattern. Breathing is one of the bodily processes that you can directly control, and it gives you access to influencing other bodily processes outside your control—including your amygdala and your PNS. We warn you not to disregard this tool, because many people are skeptical that something so simple could have such influence, and they miss out on a very beneficial strategy. By calming the defense response, you put yourself—instead of your amygdala—in the driver's seat.

Muscle-Focused Relaxation Strategies

As we noted, another method for relaxation training focuses on muscle relaxation, which also has been found to counter amygdala-based activation of the defense response, by reducing heart rate and cortisol levels (Shafir 2015). As part of the defense response, fibers in the SNS activate muscles in the body in preparation for responding. We often experience this as muscle tension, trembling, or twitching. Remember that this increased muscle tension is programmed into the human species when we are under stress, and people often feel stiff and sore because of it. Fortunately, muscle tension is another response, like breathing, that you can modify if you deliberately attend to it. Relaxing your muscles is also an effective way to reduce anxiety (Jasuja et al. 2014).

Most people don't realize that muscle tension builds as a result of amygdala activation. But if you take some time to observe yourself during a typical day, you may find that you often clench your teeth or tense up your muscles in your stomach or shoulders for no good reason. The activated amygdala may be the source of this tension. Muscle tension may increase at various times in the day, particularly if we find ourselves in stressful situations. Students taking an exam may grip their pens so tightly that it seems as if they are afraid the pen is attempting to escape! Certain areas of the body are particularly vulnerable to muscle tension, including the jaw, forehead, shoulders, back, and neck. Constant muscle tension produced due to amygdala activation uses energy and can leave you feeling exhausted as well as stiff and sore at the end of the day.

Muscle Tension Inventory

The first step in reducing muscle tension is to discover which areas of your body tend to tighten up when you're under stress. If you do a *muscle tension inventory*, you can locate the areas you need to focus on. Sit in a chair with your feet on the floor and your back comfortably erect. Hold your head up straight. Check your jaw, tongue, and lips to see if they are relaxed or tense. Consider whether your forehead is tense or frowning. Also check your shoulders to see if they are loose, low, and relaxed, or tightened up toward your ears. Is your stomach tense as if you expect someone to punch you in the gut? Check your hands and feet to see if you are clenching your fists or curling your toes. Some people even tense up their buttocks. Take a brief inventory of your entire body to see where you can find tension at this moment. Do this kind of quick inventory as you go through the day to be aware of the muscle tension you experience and notice particularly vulnerable areas. Everyone is unique and holds tension in different places.

Once you have an idea of the areas in your body that seem most likely to get tensed up, you're ready to learn to relax your muscles, paying special attention to your particular areas of tension. Tension in the muscles is a tight or strained feeling and can even be experienced as soreness. In contrast, muscle relaxation is a loose and heavy feeling in the muscles. To help you become aware of the difference between tension and relaxation in your muscles, put both hands on the table or in your lap in front of you. Now make a tight fist with your right hand, and compare it with your left hand resting in a relaxed position, to notice the difference between tension and relaxation. Next, holding the tension for at least a count of five, let your right hand relax, and drop it limply onto the table or your lap. Notice the way your right hand feels now that it is relaxed. Do you notice a difference? Also compare your two hands now that both are relaxed. Do you notice the right hand feeling more relaxed because you tensed and relaxed it? Often tensing and releasing muscles helps create a feeling of deeper relaxation in those muscles. You can tense and release muscles in other areas in order to deepen the feeling of relaxation. When you create a feeling of relaxation, you're communicating safety to the amygdala because continued muscle tension serves to maintain amygdala activation (LeDoux 2015).

Progressive Muscle Relaxation

By briefly tensing and relaxing muscles in the body, you're able to achieve a more relaxed body. In *progressive muscle relaxation* (Jacobson 1938), you focus on one muscle group at a time, tensing and relaxing those muscles, and then moving on to another muscle group. You *briefly* tense each given muscle group (hands, arms, shoulders, and so on) for a count of five or so, until you have tensed and relaxed all major muscle groups.

When you first learn progressive muscle relaxation, give yourself thirty minutes to complete the entire process of tensing and relaxing every muscle group. With practice, you can train yourself to relax your muscles more readily so that much less time is required. If you practice this process, paying more attention to muscles that need it and relaxing other muscles more quickly, you can usually achieve a fairly deep level of relaxation in just five minutes. With time, you will not need this entire procedure; you can work toward relaxing your whole body by only focusing the tense-and-relax strategy on stubborn muscle groups that you know are more difficult to relax. One person may learn that relaxing facial muscles is easy but relaxing the shoulder muscles takes more work. Another person may experience just the reverse, and find tension creeping back in to the jaw or forehead, which requires more attention. Learning to relax effectively is an individual process that you must tailor specifically to yourself, with your specific needs in mind.

You also may decide to incorporate the use of breath, imagery, mindfulness, and meditation as you relax your muscles. Although this may sound complicated or lengthy, most people find that they can streamline the process once they find the relaxation methods that work best for them. Even if you begin with the tensing approach, once you master the process of relaxing your muscles, feel free to use a tension-free approach, which is more efficient because it is quicker. The exercise presented here is a starting point; you should develop a process that reflects your own needs and is most effective at putting your body into a relaxed state (in other words, that counters amygdala-based activation in the body). You may ask someone to read this to you so that you can focus on the instructions more effectively.

Beginner's Progressive Muscle Relaxation

We recommend that you sit in a firm chair, with your back comfortably erect, your head straight, and your eyes closed or focused on a point three to five feet in front of you. Begin by focusing on your breathing, taking a few moments to practice slow, deep, diaphragmatic breathing. Aim for a rate of five or six breaths per minute. It can be helpful to think a word like "relax" or "peace" as you relax.

Now begin to focus on specific muscle groups. During this process, maintain slow, deep breathing. Begin with your hands by clenching your hands into fists for a count of five. Let go, and completely relax your hands. You can move your fingers or shake them to relax them. Let them relax in your lap, heavy, loose, and relaxed for a moment.

Next, focus your attention on your forearms. Tense your forearms by making fists again, and also tighten your forearm muscles for a count of five. Then relax your hands and arms and allow your hands to drop into your lap, heavy, loose, and relaxed. Release any tension in your forearms and feel the heaviness of the relaxed muscles for a moment.

Now focus on your upper arms. Pull your hands and forearms close to your upper arms to tense your bicep muscles in your upper arms for a count of five. Then completely loosen all the muscles in your arms and allow them to hang relaxed at your sides. Feel the weight of your relaxed hands and arms pulling your bicep muscles down into a relaxed state. Shake your arms if it helps to loosen and relax your muscles. Make sure your fingers are still relaxed.

Next, focus your attention on your feet and legs, always holding the tension for a count of five. Tense the muscles in your feet by curling your toes under. Release the tension by wiggling or stretching your toes. Moving up to your calves, push your heels down against the ground while flexing your toes and feet upward. Then relax your calves by stretching your legs out and resting your feet comfortably against the ground. Tense your thighs by pushing your feet into the ground. Then release the pressure and place your feet slightly forward in a relaxed position on the floor. Focus on the sensation of relaxation in your thighs, calves, and feet. Finally, tense your buttocks. Then release the tension and move to sit comfortably in the chair, with your legs extended slightly and relaxed. Focus on the heavy feeling of relaxation in your legs.

Now focus on your facial muscles, always tensing each set of muscles for a count of five. Start with your forehead muscles by frowning to tense them. To relax them, lift your eyebrows up, and then lower them and relax your forehead muscles into a comfortable position. Focus on the jaw, tongue, and lips by clenching your teeth together, pushing your tongue against your teeth, and pushing your lips together. Open your mouth as if yawning to release the tension, and then allow your lips to be loose and relaxed, your mouth slightly open, with your tongue relaxed. This is a good time to check to make sure that your breathing is still slow and deep.

Now, focusing on your neck, tip your head back and tense your neck. To relax your neck, bring your chin toward your chest and then roll your head to both the right and the left. Return your head to an upright position, looking straight ahead, and bring your shoulders up toward your ears to tense them and hold the tension. Then lower and relax your shoulders completely, allowing the weight of your relaxed arms and hands to pull your shoulders down. Finally focus on your torso and tighten the muscles in your abdomen as though you are preparing for a punch to the stomach. Then relax completely, allowing your stomach muscles to be loose and soft.

Take a few moments to feel the sense of deep relaxation throughout your body. If you find tension in any area, take time to tense and relax those muscles. Keep breathing in a deep and slow manner for a few moments to deepen the relaxation. Then stretch yourself comfortably and return to other activities.

Practice progressive relaxation daily, preferably at least two times a day at first, until you have reduced the amount of time it takes to relax all your muscles to approximately ten minutes. Most people learn to relax most of their muscles without tensing them first, although certain muscle groups may require stretching or tensing to achieve relaxation. Each person has muscle groups that are more likely to become tense, so the tension/relaxation process should be focused on your own needs.

If you suffer from injuries or chronic pain, tensing some muscles may be counterproductive. If so, you can follow the same guide as for progressive muscle relaxation but *without tensing muscles*, focusing on achieving relaxation in each muscle group in turn. For the most effective approach to reducing activation of the defense response and activating parasympathetic responding, combine breathing-focused methods with muscle relaxation.

Imagery-Based Relaxation Strategies

Some people, for various reasons, find that neither breathing-focused approaches nor muscle relaxation produces a relaxed body. Fortunately, using imagery, or visualization, can also create a relaxation response in the body. Some people have the ability to imagine themselves in another location and use this visualization to relax effectively. In fact, many people with OCD have a very visual, creative cortex that is capable of wonderful feats of imagination. Unfortunately, a very creative cortex can be used to imagine some very anxiety-provoking situations. If you have one of those cortexes, recognize that your imaginative ability can be used to calm your amygdala as well as activate it. As we noted in chapter 3, the thoughts and images produced in your cortex can have a strong impact on the amygdala, which does not necessarily recognize the difference between the thoughts and images that are products of the imagination and those that are due to events that are actually occurring. Fortunately, scenes you imagine can be used to calm, as well as activate, the amygdala.

The ability to visualize other scenes is a very individual ability, with some people finding imagery difficult to use, and others having no difficulty taking themselves to a pleasant location in their minds. We will help you explore whether you can use imagery to achieve relaxation. The important goal is for you to find a way to achieve *deep breathing* and *relaxed muscles*. How you get there doesn't matter as long as you achieve the deeply relaxed state that is the key to reducing amygdala activation.

To assess your ability to use imagery, read through the following description of a relaxing scene and then take five minutes or more to close your eyes and imagine yourself in that situation. Try to incorporate all your senses in the process, considering what you see and hear as well as what you feel and smell. Have someone read this paragraph to you, if that improves your visualization.

Imagine yourself on a warm, sunny beach, with blue skies dotted with several immense white clouds. Feel the warmth of the sun on your skin and the cool, misty breeze as it comes off the water. Listen to the repetitive sound of the waves as they wash against the shore. Hear the call of the seagulls in the distance. Allow yourself to relax and enjoy being at the beach for several minutes.

After giving yourself some time to imagine the scene, consider how well you were able to put yourself into the described setting. Could you hear and see the scene and incorporate your other senses as well? If you could achieve the visualization quite readily and you found it pleasant, engaging, and relaxing, we highly recommend that you use imagery as one of your relaxation strategies. On the other hand, if you found it difficult to relax using this method or found your mind wandering before five minutes were up, you might find other strategies more useful or perhaps need more practice with mindfulness before you use this approach.

If you're encouraged to try an imagery-based relaxation scene, you can ask someone to read you the next few paragraphs. Before the person begins reading, close your eyes and deepen your breathing. You can also use your own recording of the paragraphs. Remember that the goal of using imagery is to achieve relaxation. Still your body and relax your muscles as you mentally travel to the scene described below. We have provided a guided script based on a peaceful wooded scene to give you an example of the visualization process, but once you understand the method, feel free to choose any location to imagine a scene that you would enjoy. It can be a familiar one or a location you've only visited in your imagination. The key to effective visualization is to close your eyes and try to experience the scene with all your senses, including sight, sound, smell, touch, and even taste. Give yourself around ten minutes to imagine and explore the scene as it is read, with silent pauses to allow you to experience the scene. The scene is designed to give you the opportunity to include your own visual images, sounds, and smells.

Imagine yourself walking toward a beautiful forest glen, shimmering with golden and red fall colors. The air is warm, but comfortable, and the shady forest is inviting. You see several paths winding off into the trees, and you stand for a moment enjoying the fall scene and feeling the warm sun on your back. Then you choose one of the paths, and begin to enter the woods. As soon as you are among the trees, you feel the coolness of the woods and walk slowly along in the shade. You look around on the ground and see a few leaves have already fallen on your path. You glance around, looking for any signs of wildflowers. As the path comes around a corner, you hear the sound of leaves rustling ahead on the path and you stop expectantly, looking ahead to see what you might encounter on the forest path. Take a moment to look to see what is around you.

(Pause for a minute or two.)

As you continue on your way, you hear the calls of birds above you. You look up at the colored leaves in the branches above. You can see the blue sky through the trees, but what attracts your eyes are all the colorful leaves. The shapes and colors of the leaves are shifting in the breeze. You hear the sounds they make as they sway. In the branches above, you see some birds, and you stop to watch them to see if you can identify them. Take a moment to listen and watch to see all that is around you.

(Pause for a minute or two.)

As you continue on your way, you notice a stand of pines, branches rocking softly in the wind. You stop to look up at the pines, admiring them and inhaling their fresh scent. Then you hear the sound of flowing water, and look around for the source. Down a small hill you can see a little brook winding through the trees. You listen to the sound as it flows gently over and around the rocks. You approach the brook, eager to look at the water and perhaps touch it. Take a moment to enjoy the brook.

(Pause for a minute or two.)

This time in the woods has been a special and peaceful time. You see the pathway winding out of the shade into the sun ahead of you. You can't help but stop a minute to enjoy the sights, sounds, and smells of the woods for a few more moments. Looking around, you are trying to remember this moment, this scene. You are cherishing the peacefulness of this time and place. Take a moment one last time to see what is around you.

(Pause for a minute or two.)

Now, gradually come back to your current place. Move your arms and legs a bit to be more conscious of your body. Become more aware of the sounds around you. Open your eyes to look at what is around you. Welcome back!

Hopefully, after visualizing this scene, you return to the world with a calmer amygdala! As we noted in chapter 3, the amygdala is strongly affected by the thoughts and images that are being produced in the cortex. If you can successfully calm yourself by imagining being in a relaxation-producing situation, make use of that to reduce the activation of the amygdala. If you're encouraged to try imagery-based relaxation, you can come up with your own scenes or find many available on a variety of apps on your phone.

Meditation Practices

Meditation is another way of achieving relaxation. Meditation is the process of focusing attention, perhaps on the breath, perhaps on a specific object or thought. Various types of meditation—including mindfulness, an approach that is growing in popularity—have been shown to have effects on the amygdala (Goldin and Gross 2010). Studies show that certain meditative practices, including mindful attention to breath (Doll et al. 2016) and awareness-based compassion meditation (Leung et al. 2018), reduce activation in the amygdala. Mindfulness has been shown not only to affect the amygdala, but also to have measurable effects in the cortex (Davidson and Begley 2012; Taren et al. 2013). Because meditation affects the cortex as well as the amygdala, we'll provide a more detailed explanation of mindfulness in particular in chapter 11 as well as in the bonus chapter "The Healthy Use of Worry," available at http://www .newharbinger.com/47186. Here we'll discuss *breath-focused meditation* because it is a very effective method of calming the amygdala (Doll et al. 2016).

You may already be interested or experienced in meditation, and if so, we strongly encourage you to continue this pursuit. Research has shown that engaging in meditation regularly can reduce a variety of stress-related difficulties, including high blood pressure, anxiety, panic, and insomnia (Walsh and Shapiro 2006). In the amygdala, which is the focus of this chapter, meditation has been shown to have direct and immediate calming effects, even for beginning meditators (Taylor et al. 2011). Both short- and long-term effects occur in meditation. During meditation, the activation in your body shifts from a sympathetic dominant to a parasympathetic dominant state (Jerath et al. 2006), meaning that the FFF response gets shut off. Researchers have also found that, after learning meditation strategies, when people encountered images that produced negative emotions, their amygdala activation was reduced (Doll et al. 2016; Leung et al. 2018). Additionally, mindfulness has been shown to be associated with lasting changes in the connections between the amygdala and other parts of the brain (Taren et al. 2013), which is thought to help provide calming information to the amygdala. Many people find that incorporating regular meditation into their morning routine decreases their overall anxiety for the day. Meditation has been shown to be a beneficial treatment for those with OCD, not only by reducing anxiety that results from amygdala activation, but also by decreasing OCD symptoms, including intrusive and obsessive thoughts (Rupp et al. 2019). After mindfulness training, the connectivity is reduced in the frontal lobe circuitry that is thought to promote obsessive thinking, and increased in areas that promote present moment awareness (Yang et al. 2016). These findings should encourage you to see whether meditation can be of benefit to you.

For centuries, many meditative practices have focused on the breath, with the goal of concentrating on the experience of breathing or modifying breathing in some way. One meditative practice that people with OCD have found particularly helpful is attention to the breath, especially a technique known as the three-minute breathing space (Hertenstein et al. 2012). This approach, which can be easily found on YouTube, emphasizes shifting your focus from your current experience to your breath, and then returning to your current experience with a changed awareness.

Using meditation effectively requires some practice. In most studies, people received at least sixteen hours of training prior to being assessed as to whether practicing meditation had changed their amygdala functioning, and others had weeks of training. So, for maximum benefits, you may want to seek training from an instructor or a therapist. Resources are also plentiful on the internet, and a variety of apps are also available to assist you in practice.

Meditation techniques that focus on breathing have demonstrated effectiveness in modifying the amygdala's response. In addition to reducing amygdala activation (Doll et al. 2016), they result in the activation of a parasympathetic state (Jerath et al. 2014; Jerath et al. 2015). When you try the following exercise, remember that the focus on your breath can calm both your amygdala and your body.

Sit in a comfortable seat that allows you to place your feet on the floor. Either close your eyes or focus your gaze on some point five to ten feet in front of you. Rest your hands in your lap or at your sides. Now simply focus your attention on your breath. Breathe through your nose, and notice the way that the air feels as it travels through your nostrils. Don't force your breath; just take in long, slow, deep breaths and observe the different sensations of inhaling and exhaling, first in your nose and then in your chest. Enjoy the sensations of breathing. You may inhale or exhale through your mouth or your nose. Feel free to choose what seems best for you. If you breathe through your mouth, notice the sensations in your mouth, as well as your tongue, lips, and jaw. Are you clenching your jaw? Are you pressing your lips together or relaxing them?

Notice any differences between the air that comes into your nostrils or mouth and the air that is leaving. Pay attention to the way that the inhaled air expands your chest as your lungs expand. Observe the different stages of breathing: As you inhale, notice the feeling as air fills your lungs, and as you exhale, notice the feeling as your lungs empty. Then focus only on the process of inhalation and how it feels when you begin the process of inhalation and how it feels different as you continue until the lungs feel full. Note the changes in the process of exhaling: How does exhaling feel at the beginning, middle, and end? Keep your breath slow and deep, and notice if you hold your breath at any time. Try not to hold your breath.

During this meditation, your mind will probably wander to other sounds or thoughts. This is natural and to be expected. When your mind wanders, just bring your focus back to the sounds and sensations of your breath. It is normal for your mind to wander multiple times; just bring it back thirty or forty times if you have to. You are training yourself to focus attention, and bringing your attention back strengthens your ability to focus, just like building a muscle requires practice.

If you want to incorporate imagery, you can imagine that you are breathing out all the stress and anxiety in your system, and filling yourself up with clear, clean air. You can imagine your stress being a certain color, and imagine breathing that color out with each breath, and breathing in clear air, until you feel you have emptied out the stress and filled yourself up with clarity.

Or you can think of a word or phrase as you breathe, in order to associate that word with the feeling of relaxed, deep breathing. You can think "relax" or "calm" or a phrase like "filled with peace."

After you have spent about five minutes focusing on your breath, slowly and gently come out of the meditation, and return to your day. Hopefully, you will notice a change in your stress level and your body after you take this time to focus on your breath.

We've covered several approaches in this chapter to calming the amygdala, combatting the defense response, and reducing anxiety: breathing-focused relaxation, muscle-focused relaxation, image-based relaxation, and meditation. There is no single right way to achieve the relaxation that can reduce amygdala-based anxiety; you simply need to find which techniques work best for you. But the ability to relax is only beneficial if you use it routinely, so make sure you choose strategies that you can incorporate into your daily life. If you include relaxation as a part of your morning or evening activities, or incorporate it during your work breaks or while using public transportation, you're more likely to establish it as a part of your daily routine. Deliberately scheduling two or three opportunities for brief relaxation breaks during your day will keep your amygdala calmer and your body in a state where the parasympathetic response is dominating. That will mean that you're more likely to stay in control rather than be under the control of your OCD as the day goes on. Next we'll explore other key elements in calming the amygdala: exercise and sleep.

CHAPTER 7

Exercise and Sleep
for the Amygdala

In this chapter, you'll learn strategies to care for the amygdala in a way that keeps it calm and makes it less likely to activate the defense response. As you have seen, an activated amygdala means that you experience the physical aspects of the defense response as well as the cortex's response to all of those physical changes. Anxiety is often a result of the defense response. A calm amygdala means you're more likely to have a relaxed body, calmer thoughts in your cortex, and less anxiety. Exercising and getting healthy, extended sleep are essential for having a calm amygdala. Exercise in particular has surprisingly powerful effects in the brain, rivaling the effects of medications in helping you rewire your brain.

Exercise: What the Amygdala Wants

Remember that one purpose of the defense response is to get the body ready to fight or flee from danger, so some kind of physical activity is exactly what the amygdala wants the body to engage in. When it initiates the defense response, the amygdala activates your sympathetic nervous system, preparing your body for physical exertion. And when you engage in physical activities, especially aerobic ones, amygdala activation tends to decrease. It's as if you have done what the amygdala wants, and it can now stand down.

You may have been advised that exercise will help you manage your anxiety but didn't see the logic of it. If you're worrying about flaws in an upcoming presentation you must give, what good will it do to go for a run? Or if you have fears of contamination that lead you to feel panicky, how can getting on the exercise bike help? It doesn't make sense until you understand how the amygdala works. When the amygdala is activated and producing the defense response, your body is prepared to act, so you should! And when you exercise, your anxiety will decrease, even if the situation you think is producing your anxiety hasn't changed. It's not logical, but it is fact. Over the past twenty-five years or so, the majority of studies on exercise have shown that it reduces anxiety (Ensari et al. 2015; Rebar et al. 2015). Just twenty minutes or less of exercise can result in less anxiety (Anderson and Shivakumar 2013; Chen et al. 2019).

We recommend that you consider exercise as an amygdala-calming response whenever you're feeling anxious, stressed, or panicky. Don't get caught up in looking for logical reasons why exercise matters in the specific situation. Once, when Catherine's college-age daughter was on the phone, explaining problems in her relationship with a boyfriend, she said, "Mom, I'm just going to stop talking to you about these problems, because it's making me more anxious. I'm just going to go for a run and I will feel better." Whether you're anxious about an upcoming exam, fearing you have offended someone, or trying not to panic because you had thoughts of harming your child, exercise will result in less activation in your amygdala in a matter of minutes.

How does this work? If your amygdala has activated the defense response, exercise makes use of the increased blood flow to your extremities and the increased heart rate and blood pressure, reduces your muscle tension, burns off the excess adrenaline in your system, and stimulates the production of endorphins (Bourne, Brownstein, and Garano 2004). Exercise also increases brain levels of serotonin (Johnsgard 2004), which is what many medications prescribed for anxiety do, so people who want to avoid medication can try exercise (Greenwood et al. 2012). Aerobic exercise that involves large muscle groups seems to be most helpful in reducing amygdala activation (DeBoer et al. 2012). This includes running, walking, cycling, swimming, rowing, aerobic dance, and jumping rope. Brisk walking is perhaps the exercise most accessible to the most people, and using a treadmill certainly counts.

Also, you'll benefit more if you begin a regular exercise program. When participants *regularly* exercised at least twenty-five minutes every two or three days for a total of twelve exercise sessions, they began to experience *less anxiety in general* than those who were not exercising (Lattari et al. 2018). People with high anxiety who begin a regular exercise program find that, as the program goes on, their anxiety decreases more and more after exercise sessions (Lucibello, Parker, and Heisz 2019). Studies show that exercise programs reduce anxiety even in older adults (Mochcovitch et al. 2016). Exercise does not require training; brisk walking can result in changes in your anxiety.

Research shows that regular exercise results in changes in the amygdala itself. After regular exercise, changes occur in the amygdala in rats and mice (Greenwood et al. 2012; Kim et al. 2015) as well as in humans (Bernacer et al. 2019). (We can't always investigate changes in the human amygdala because some investigations require dissection, but the amygdala seems to operate in very similar ways in different mammals.) Animal studies show that exercise appears to affect a certain type of neuron found in the amygdala's lateral nucleus, the part of the amygdala that decides whether a situation is dangerous (Greenwood et al. 2012), and regular exercise makes these neurons less active, resulting in a calmer amygdala that is less likely to initiate a defense response (Heisler et al. 2007). One human study showed that exercise changes the connectivity between the amygdala and the anterior cingulate cortex (Bernacer et al. 2019), areas that are both connected to difficulties seen in OCD. Another human study showed changes in amygdala reactivity after only fifteen minutes of exercise (Schmitt et al. 2020). These changes are correlated with changed connections between the amygdala and the insula (a part of the brain involved with experiencing emotions), which, in turn, are associated with an improved mood and reduced fear.

As you can see, other parts of the brain besides the amygdala are affected by exercise. Research confirms that exercise promotes growth of neurons both in rats (DeBoer et al. 2012) and in humans (Schmolesky, Webb, and Hansen 2013), strengthening the evidence that exercising can promote changes in the brain, including in the cortex. Exercising can increase levels of neurotransmitters and promote the growth of new cells in the cortex (Cotman and Berchtold 2002). Extended or intense work-outs also cause the release of endorphins into the bloodstream, which not only produce a feeling of exhilaration but also affect activity in the cortex and help combat depression.

Exercise also affects compulsions. Because we engage in compulsions to reduce anxiety, anxiety fuels compulsions. If we use exercise to reduce anxiety, we reduce the fuel for compulsions, making it much easier to resist them. If you've become accustomed to using some compulsive behav-ior when you feel dread, anxiety, or panic, you can replace that compulsive behavior with exercise. Instead of repeating your compulsions, which serves to strengthen them, remember that anxiety results from your defense response, and react to anxiety in a way that has *lasting* calming effects on the amygdala. Compulsions reduce the anxiety for a short period of time, but the anxiety always comes back because you have not chal-lenged the thoughts that lead to the compulsions or changed the amygdala in any way. Using exercise has direct calming effects on the amygdala, leading to less anxiety, making it more beneficial than compulsions. *Regular* exercise results in an amygdala that is generally calmer, so you experience a lower level of anxiety in your daily life. That also means less fuel for your compulsions.

The best kind of exercise for you is exercise that is moderately intense, approved by your doctor, and enjoyable to you. Having it be enjoyable is important, because it means that you're more likely to do it. It's also helpful to choose an exercise that a friend or family member will join you in, because when you exercise with someone on a regular basis it forms a com-mitment that keeps you engaged. You can choose more than one type of exercise, walking in the evenings and biking on the weekend, for example, in order to keep your interest up. *The goal is for you to engage in exercise a minimum of three times a week for at least thirty minutes.* As long as the exer-cise gets your heart pumping so that your heart rate is up, you're doing exactly what the amygdala wants (using your sympathetic nervous system and burning off adrenaline), reducing anxiety both immediately and in the long term, and making changes in your amygdala that make it more resistant to anxiety. Watch what happens with your anxiety when you increase your level of exercise. When you recognize a reduction in your anxiety, improvement in your mood, and more control of your OCD, you will be motivated to stick with your program!

Sleep: An Aid for Your Amygdala

Although our culture minimizes the importance of sleep, if you have OCD, you cannot afford to disregard the impact of poor sleep on your amygdala. When you have OCD, you are often driven by perfectionistic thoughts, the need to complete tasks in a particular (correct) way, or the habit of overthinking, so that time frequently gets away from you and sleep becomes a casualty. Cutting back on your sleep while trying to get something to be "just right" may seem essential when you have OCD. It's easy to make sleep less of a priority without even realizing it. People tend to think of sleep as a period during which the brain shuts down, but actually the brain is carrying out some very important tasks while you sleep. As you sleep, your brain is busy making sure that hormones are released, needed neurochemicals are produced, toxins are removed, and memories are stored. If you don't get enough sleep, these tasks won't be accomplished. But even more importantly, when deprived of sleep, your amygdala is more easily activated, creating increased anxiety that worsens OCD. (We should note that poor sleep can be due to multiple causes, including pain, hormone imbalances, breathing difficulties, and a variety of disorders, which may require evaluation and treatment by a sleep specialist. Please seek these if necessary.)

Research has revealed that after someone is deprived of sleep, the amygdala is more reactive. When college students were asked to sleep less before showing up for an experiment, and they were compared to other college students who had adequate sleep the night before, clear differences in amygdala reactivity were demonstrated. The amygdalae of sleep-deprived students reacted more strongly to emotional images (Yoo et al. 2007). When you consider periods of your life when you were sleep deprived, was your anxiety worse? We often see this with parents of infants who find their anxiety skyrockets during the first months when they're very sleep deprived. In addition, college students who are very sleep deprived during midterms or final exams often have difficulty managing their anxiety. Sometimes the amygdala produces irritability as well as anxiety, reflecting the fight as well as flight response. But people seldom consider the detrimental impact of limited sleep on the amygdala, because research has only recently uncovered the connection.

Sleep Difficulties and OCD

When people have an anxiety-based disorder like OCD, simply telling them to get more sleep is not sufficient. Just because you want to get more sleep doesn't mean that you can. This is because anxiety difficulties, including worry and rumination, have been shown to be associated with sleep difficulties (Spoormaker and van den Bout 2005). The amygdala may be at the root of these difficulties. Perhaps our ancient ancestors who had an amygdala that kept them alert and awake during times of danger were more likely to survive. Whether a roving pride of lions had been spotted or an earthquake had recently occurred, those with an amygdala that remained activated and interfered with their sleep would be more likely to be alert and responsive to danger. They and their offspring were more likely to survive. We're likely to be the descendants of those who had an amygdala that interfered with sleep when danger appeared to be present. Although it's difficult to scientifically demonstrate the development of this tendency in our ancestors, it makes sense that it is an evolutionary advantage to have an amygdala that keeps you awake if a potential threat exists.

In today's world, the stresses that we face are not likely to be ones that staying up all night will protect us from. When we can't pay the mortgage or we find out that a parent has cancer, staying up all night won't protect us. But this is still how the amygdala seems to respond. Whether the danger is your worry that someone will find grammatical errors in your writing or the nagging doubt that you ran over a dog without realizing it, if you begin considering those kinds of thoughts as you're lying in bed, your amygdala is likely to become activated and sleep will be difficult.

Avoiding Amygdala Activation Before Bedtime

The first goal to have in mind in order to fall asleep is to make sure your amygdala is not activated before your bedtime. This means that you should have a period before bedtime during which your thoughts are not focused on threats or worries. Often we're effective at keeping busy during the day so that our minds aren't focused on our worries, but the minute we stop to rest and clear our heads, worries jump into our minds! This is completely normal because concerns are likely to be brought to your attention in a quiet period when you're not focused on other responsibilities and situations. But before bedtime is not the time of day to focus on your concerns and worries. We should focus on them at a more opportune time, but often we avoid them. For example, if you put off sitting down to do the bills all day—perhaps to avoid the distress of examining what funds are available or deciding which payment you need to put off—you're likely to be feeling very stressed just before bedtime. You need to schedule stressful activities or decisions earlier in the day or evening as much as possible. For some people, this will even include scheduling time to worry or obsess at a period during the day that will not interfere with sleep.

As you have learned, telling yourself *not* to have certain thoughts is not effective. That approach guarantees that you're more likely to think about those thoughts. A more effective method is to tell yourself *what else* to think about, in order to replace the worry-focused thoughts. This means that you need to find something calming or interesting to focus your attention on before you fall asleep, either before or after you get into bed. You may choose to focus on breathing or other relaxation strategies after you get into bed in order to replace bedtime worrying, if you find those strategies are effective. One warning, however, is that you can't expect to become immediately relaxed after you've spent a half hour focused on something stressful (like obsessing over whether the puzzled look your boss gave you means she's not going to promote you). Focusing on thoughts and worries that activate the amygdala in the hour before retiring to bed is likely to create amygdala activation that can interfere with falling asleep.

Some people say that listening to music allows them to fall asleep effectively, and if that works for you, great! But others worry right along with the music because, typically, people are able to listen to music and think other thoughts at the same time. Many people say that reading helps them fall asleep, and this can be effective. It's difficult to worry at the same time that we're reading, because both reading and worrying are more thought-based and verbal, so reading interferes with the thinking process needed to generate worry. If you choose to read before going to sleep, remember that the goal is not to delve into an exciting novel that you'll stay up all night reading. The goal is to read something that will calm and relax you, something you can willingly put aside once you become sleepy.

Having someone talk to you can very effectively interfere with worry because it's difficult to think about anything (including worry) when someone is speaking to you. This is a great example of how it's impossible to focus on two tasks at the same time. You can take advantage of this inability to focus on more than one thing in order to keep your amygdala calm.

If you listen to an audiobook, a calming podcast, or another recorded program, you can pretty effectively block the worry circuits in your cortex from taking over your focus. Of course, you need to focus on what you're listening to because if you tune it out and focus on your own thoughts, you'll be back on the Worry Channel. You can think of this process as making sure that you're tuning your amygdala into a channel that will not activate it. If your amygdala is listening to a Charles Dickens novel or a podcast about selecting plants that will bloom all year-round, it will react with a big "Yawn! Nothing to see here." Compare that to how the amygdala will react if you're worrying about whether your sore throat means you're going to need your tonsils removed. Or what if you're considering whether your boyfriend's delay in calling means that a breakup is imminent? In summary, you need to approach your bedtime with the clear intention to keep your amygdala from becoming activated by worries or distressing thoughts.

What Kind of Sleep Does My Amygdala Need?

Sleep occurs in stages that are characterized by different types of brain waves and different activities occurring in the brain. We cycle through these stages repetitively, in a programmed manner, with one stage, called rapid eye movement (REM) sleep, occurring several times during the course of the night. (See figure 7.) REM sleep is the stage during which we have our most vivid and memorable dreams. By deliberately waking individuals to deprive them of specific stages of sleep and examining whether the loss of a certain stage of sleep is more detrimental, researchers have identified that REM sleep is very important for the amygdala. The amygdala is less reactive when a person is getting plenty of REM sleep (van der Helm et al. 2011). Therefore, if you want to keep your amygdala calm, you need to make sure that you get enough of this specific kind of sleep.

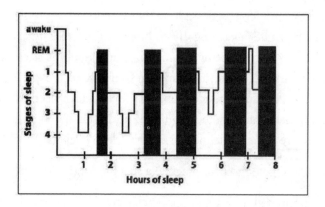

Figure 7

To get a sufficient amount of REM sleep, you need to understand exactly when REM sleep occurs in the sleep cycle. In figure 7, you'll notice that REM sleep (shown by the black vertical bars) does not occur until you go through the first complete sleep cycle, which typically takes at least an hour. REM sleep periods are short at first and tend to get longer as you go through each sleep cycle. This means that in order to get long periods of REM sleep, you need to sleep at least seven to nine hours. Also, the sleep cycle follows a certain pattern, and if you wake up during the night and are up for even twenty minutes, that's often enough to disrupt the sleep cycle. Bonnet (1985) found that periodic, brief disruptions of a person's sleep caused an altered distribution of sleep stages. For example, when individuals' sleep is disrupted by bruxism, or grinding of teeth, sleep stages were altered and REM sleep was fragmented (Kishi et al. 2020). Research shows decreased amygdala reactivity is associated with better consolidated REM sleep (Wassing et al. 2019) and with specific physiological reactions that occur in the brain during REM sleep (van der Helm et al. 2011). Overall, evidence indicates the importance of REM sleep in reducing amygdala activation.

In short, if you want to have the most REM sleep you can get in order to have a calmer amygdala, you need to sleep continuously for as long as possible. Go to bed early or sleep later, whatever works to give you an extended period of uninterrupted sleep. If you wake up in the night, get back to sleep as quickly as possible so that you don't have to start the sleep cycle over from the beginning. If you practice *good sleep hygiene*, or habits that are effective in initiating, protecting, and extending sleep, we believe you'll find yourself enjoying a decrease in your anxiety. People who become more aware of the importance of extended periods of sleep in controlling anxiety often become more committed to better management of their sleep. Here are some guidelines for healthy sleep hygiene:

- Before going to bed, practice a specific routine that's relaxing.

- Eliminate light stimulation, including screens, for at least an hour before bed.

- Exercise during the day, avoiding the hours before sleep.

- Avoid napping, or limit naps to "power naps" of no more than fifteen to twenty minutes.

- Establish a consistent bedtime and waking time to give your brain a stable routine.

- If worries or obsessions trouble you at bedtime, set a worry or obsession time during the day.

- Near bedtime, replace amygdala-activating thoughts with relaxing ones.

- Replace thoughts with reading or listening to podcasts or audiobooks, if necessary.

- Ensure that your sleeping environment is conducive to sleep (e.g., dark, cool).

- Avoid caffeine, alcohol, and spicy foods in the late afternoon and evening.

- If you can't fall asleep after thirty minutes, get up and do something relaxing in a dark environment for an hour.

- Make sure your bed is comfortable.

- Use your bed primarily for sleep.

- Avoid and limit use of sleep aids because they only work temporarily.

In summary, if you take care of your amygdala well by getting plenty of aerobic exercise and REM sleep, you'll have a much calmer amygdala than you did before you knew how important these two aspects of your life are to the amygdala. While it's not easy to commit yourself to lengthy sleep and regular exercise, and circumstances will certainly interfere with doing so during certain periods of your life, you're much better off knowing that you can use these strategies to keep your amygdala calmer. You greatly improve your ability to direct your life without the restrictive influence of OCD when you reduce anxiety by cutting back on the amount of time you have an activated amygdala. The next step is going beyond simply reducing the activation of the amygdala, and learning how to teach the amygdala to respond differently. We'll turn to this in the next two chapters.

CHAPTER 8

The Language of the Amygdala

In this chapter, you'll learn the language of the amygdala to understand how it knows what's dangerous, how it learns, and how it communicates to you. First, let's take a moment to review what you've already learned about the amygdala. The amygdala's role is to serve as your protector; the way it's connected into your brain, it serves rather like an alarm or early warning system that can create a response before you even consciously know a danger exists. But the information the amygdala receives may not always be detailed or complete enough to make correct decisions. The amygdala is very capable of responding as if danger exists even when none does. Despite its fallibility, the amygdala is the part of your brain that decides whether your body will react to an object or a situation as if it is dangerous. It creates a defense response (the FFF response) that can prepare your body to respond to a threat in milliseconds. (If you need to review this information, see chapters 2 and 5.)

How the Amygdala Knows What's Dangerous

If the human amygdala is supposed to help us detect danger, how does it know what is dangerous and what is not? First, the amygdala is prewired to respond to some objects and situations as if they are dangerous (Ohman and Mineka 2001). The amygdala seems to automatically see insects, animals, heights, water, angry facial expressions, and contamination as threatening, because humans learn to fear them with very little prompting. Why should children fear tiny spiders or ants and seem to have little fear of automobiles, when automobiles are much more harmful? Most likely, fears of these kinds are built into the human amygdala because they have given us a survival advantage through our evolutionary history. So concerns about contamination, illness, and blood and other body fluids, common in people with OCD, may have an evolutionary explanation. We should note, however, that even fears that appear to be programmed into the amygdala can be changed. If they couldn't, we wouldn't see people disregarding fears of heights as they enjoy zip lining in the mountains or bungie jumping off cliffs. Or how about those of us who sleep with sharp-toothed little animals rather than fear them?

On the other hand, the amygdala also *learns* to see certain objects or situations as threats on the basis of your life experiences. The amygdala is constantly learning. After a negative experience, the amygdala creates brain circuitry that causes a person to fear a previously unfeared object. Consider little Johnny, who was knocked down by a child who was unsteadily riding a bicycle down the sidewalk. Since then, every time Johnny sees a bicycle, he shows signs of fear and panic. He doesn't want to touch any tricycles or bicycles. This is a result of the amygdala. To protect Johnny, the amygdala creates memories that identify any object associated with negative events as a threat. When it detects an object similar to those stored in memory as a threat, the amygdala creates the defense response. This allows the amygdala to create specialized neural circuitry that helps people be alert for the specific dangers that occur in their own lives in order to avoid them. This powerful advantage has kept the amygdala useful and virtually unchanged for millions of years. Some obsessions are based on this same learning process.

You also may have noticed that even though Johnny had a negative experience and his amygdala now sees bicycles as a threat, the amygdala has not made an accurate memory. Every bicycle that comes down his sidewalk is not going to hit him, and touching a bicycle will not hurt Johnny. This is a common problem with the amygdala. After something negative occurs in a person's life, the amygdala can learn to treat certain objects or situations as if they are threatening (based on the very strong memories that it stores). This response can last for years.

Your amygdala creates lasting memories about what may be dangerous based on your own unique history. The memory is not a "recording" of what happened; the memory simply associates something with danger. After it is formed, the memory will cause the amygdala to respond to this object or situation as if it were dangerous. After a situation or an object is tagged as a threat, the amygdala will be vigilant for this situation or object, produce a defense response whenever you encounter it, and also focus attention on it in the cortex in a way that can promote obsessive thinking.

Note that the amygdala can see something as a threat even if the person does not remember a negative experience that caused the fear. After several years, Johnny may not remember being knocked down by a bicycle, because the memory may not be retained in his cortex, but the amygdala is fully capable of holding on to the memory of bicycles being a threat for years and years. Amygdala-based memories can be very strong. Because the memory is from the amygdala and not from the cortex, Johnny can keep his fear of bicycles even without a memory of how it developed. In a similar way, your obsessions may be influenced by your amygdala, even though you may not recall a negative event associated with the situation.

Genetics can also influence the amygdala and its reactions, not just in people but also in other animals. You can inherit an amygdala that creates more defense reactions than the amygdalae of other people. One study found that children who have a smaller left amygdala tend to have more anxiety difficulties than other children (Milham et al. 2005). But, whether you have inherited a more sensitive amygdala or not, your amygdala is capable of learning to respond differently if you know how to teach it.

How the Amygdala Learns

We have noted that the amygdala learns through experience. As you live your life, the amygdala is learning *through your specific experiences* what is dangerous and what is safe, what is distasteful and what is pleasurable. The amygdala learns through the *process of association*. You can learn to experience something that is perfectly safe as terrifying if it is paired with a negative experience. Perhaps when you were a toddler you went to McDonalds and were given some chicken nuggets with a little barbecue sauce container. Instead of dipping the nuggets in the barbecue sauce, you put both hands in the barbecue sauce and then, for good measure, spread the sauce on your face. When your parents saw this, they immediately shouted and grabbed napkins to clean you up. If their emotional reaction was a negative experience for you, your amygdala would have learned from this experience. When a negative experience occurs, such as being shouted at, it creates excitation and activation in the amygdala. Neurons in your little toddler amygdala are firing. This excitation is occurring at the same time that the amygdala is also processing other sensory information, such as the sight and feeling of barbecue sauce on your fingers. Because barbecue sauce on your fingers is being *processed immediately before or at the same time as an activating negative experience,* barbecue sauce on your fingers is tagged as a threat. The amygdala creates a negative emotional memory about having your fingers coated with something red. If you had been tasting the barbecue sauce, the taste of the sauce may even become something that distresses you.

This emotional learning, based on associations, occurs when an object or a situation is paired with a negative event. (See figure 8.) Some readers may recognize that this kind of learning is called classical conditioning. In studies on rats, scientists have actually observed that connections form in the amygdala when such pairings are experienced (Quirk, Repa, and LeDoux 1995). The amygdala is creating emotional memories when something is paired with an emotionally relevant experience. This is how we learn to fear our father's angry voice (he yells before he punishes us), to feel happy to see our grandmother (she always shows up with treats or presents), or to fear driving on icy streets (after sliding into a parked car one morning). Some experiences can be positive, as in the case with grandma; however, we are more concerned with the negative experiences because they teach the amygdala to produce the defense reaction that underlies the anxiety that fuels OCD.

Association/Pairing

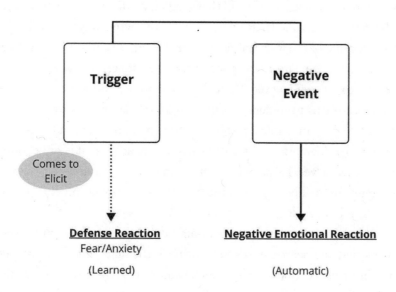

Figure 8

The association-based language of the amygdala is what creates many of our emotional reactions. In the case of anxiety, the amygdala is creating memories that identify something as a threat. After this connection has been created, the sights, sounds, or smells associated with the negative event will become capable of activating the amygdala. Whenever the amygdala recognizes situations similar to what has been tagged as a threat, it will create the defense response. In the example of your visit to McDonalds, many different sensations associated with the barbecue sauce could now activate the amygdala. Not just the taste or smell of barbecue sauce but also the sight of something red on your fingers could cause the amygdala to create the defense response, and your parents now have a toddler who becomes distressed when a bit of ketchup gets on their fingers.

Depending on the situation, almost anything can become a trigger. The term *trigger* refers to an event, an object, a sound, a smell, a sensation, and so on, that activates the amygdala as a result of previous association-based learning. A trigger can be the smell of a cologne, piles of dirty laundry, a brown smudge on your hand, or a certain country song. Triggers are an important aspect of the language of the amygdala. For an object or a situation to become a trigger, all that's required is for it to be experienced at about the same time that a negative event is activating the amygdala. This is because of the way that neurons in the brain learn—memories are formed when two neurons are firing at the same time.

Sorry to get a bit technical here, but if you want to know how to rewire the brain, it helps to understand the way that it learns. As we noted in chapter 4, the brain is filled with cells called neurons that send messages to each other. Memories are formed when the neurons build connections between themselves. Remember, neurons that fire together wire together. This statement will point you in the direction of how you can change the wiring in your amygdala.

Basically, for the neurons in your amygdala to build new connections or memories, one neuron needs to be firing at the same time that another neuron is firing. A connection between neurons is strengthened when two neurons fire together, and eventually a pattern of circuitry develops in which activation of one neuron causes the other to activate as well. Changing neural circuitry (i.e., learning) involves changing the activation patterns in the brain so that new connections develop and new circuitry forms.

Returning to the idea that the amygdala creates certain triggers, you can now understand how triggers are created through pairings: one neuron responding to the trigger is firing at the same time another neuron reacting to a negative event is firing, causing a connection to be made between the two. Remember, an object or a situation itself need not be harmful or threatening for it to be tagged by the amygdala as a threat. Any object, even a harmless teddy bear, could come to cause anxiety through association-based learning in the amygdala if it is experienced at about the same time that some threatening event is activating the amygdala. For example, when Alphonso was in the Air Force, he once received a demerit after a room inspection, something he prided himself on always passing with flying colors. He was so upset with this demerit that he sought out the captain for an explanation. She said simply, "Check the seal in your refrigerator." When he did so, he saw that the seal had a small dark mark on it. After this experience, he became obsessed with cleaning his refrigerator, and especially the seal, before each inspection. The smallest indication of a stain anywhere in the refrigerator caused him great distress. Obviously the dark mark did not pose a danger, and neither did the stains, but they had been associated with a negative experience, so they now caused Alphonso's amygdala to create the defense response. Alphonso was distressed by a stain in his refrigerator even when an inspection was not scheduled.

The amygdala's reaction to triggers can range from relatively weak to very strong depending on the aversiveness of the experience associated with the trigger. For example, a certain type of candy might be a mild trigger because it was associated with your childhood visits to the pediatrician. On the other hand, if you once ate egg salad when you had a serious illness that caused vomiting, you might find that, even years later, egg salad is a strong trigger, and even the sight of it makes you nauseous.

How the Amygdala Communicates

We have described the amygdala as an alarm, sending you a signal that you are in potential danger. But the amygdala does not communicate danger to you in words, thoughts, or images; it communicates to you through physical reactions it initiates in your body and the emotions you experience as a result. You *feel* what the amygdala communicates rather than *think* it. When the amygdala is producing the defense response, and you are experiencing the effects of the FFF response in your body, you can feel how your heart rate or breathing seems different, and you feel tense. You can also feel emotions, typically dread, distress, fear, or anxiety. Let's separate out what comes from your amygdala's influences and what comes from the cortex so that you become more aware of the language of the amygdala itself.

We'll start with the amygdala. As we said, the amygdala does not tell you what the danger is in words, but you often have a sense of what the danger is. Your attention may be focused on a particular object or sound, whether you want to focus on it or not. You may feel overwhelmed and frozen. You may feel you need to take action—for example, get away from a dog, wash your hands, or put objects in order. You experience a sense of urgency and distress as well as physical sensations: your heart may be pounding; your mouth may become dry; you may feel nauseous; you may feel a rush of adrenaline. These emotional and physical experiences are part of the language of the amygdala as it produces the defense response. The strength of the reaction you feel can vary, from a general reluctance to touch something to full-blown panic or wanting to run away.

Your cortex is the part of the brain that tries to identify the sensations and feelings you're having. It tries to put your amygdala-based experience into words, to explain or describe it. But thoughts and words from the cortex may not capture what is really going on. You don't want to tell your friend, "I'm afraid to ride in the car with you" because even though that's the feeling produced by your amygdala, another part of your brain knows she is a safe driver and the fear does not make logical sense. If you say, "I'm afraid the dog will bite me," that may not be logical or correct. You're simply feeling afraid of the dog and sensing your body react to potential danger. Even if you are reassured that the dog has never bitten anyone, the defense response from the amygdala does not go away. The cortex-based explanations don't fit very well because your reactions are based in the amygdala, not in the cortex. Expecting the cortex to know what is really happening is rather unrealistic. It would make as much sense to look in your refrigerator for an explanation about why your car will not start!

The amygdala is not logical, and the triggers that it responds to do not always pose a true danger. When your cortex tries to interpret your feelings and impulses, it can come up with explanations that seem plausible or logical, but the amygdala could be responding in error. The neighbor's dog does not pose a danger; it is simply a trigger that provokes the defense response. But that does not mean that your physical and emotional reactions are not real. You are experiencing a very real defense response. The feeling that the amygdala is producing may have been learned in a previous experience you had, and may not apply to this particular situation, but the feeling is still very real and unlikely to change in response to logic or reasoning (LeDoux 2015). The amygdala, in its role of protector, is alerting you to potential danger based on what it has learned, and while you can consider the amygdala to have "good intentions," you shouldn't necessarily trust that it's an accurate predictor. You can *feel* that you are in danger without *being* in danger.

As the cortex tries to make sense of the feelings produced by the amygdala using logical explanations, it can actually make the amygdala more activated. For example, your cortex may come up with a logical explanation for why you don't want to ride in a car with a friend: "I'm afraid to ride in the car, not because I don't trust you as a driver, but because I can't predict what *other* drivers will do." In reality, the fear coming from the amygdala is not based on this kind of logical assessment of the situation. Because the amygdala learns on the basis of pairings, your fear is probably due to a negative experience that made *being a passenger in a car* a trigger for fear. But your cortex's proposal that other drivers pose a danger, and that your friend's safe driving may not protect you, calls up images (in the cortex) of drunk drivers ramming into the car. So the cortex can unknowingly add to amygdala activation, just by incorrectly explaining the situation.

The recognition that you can't trust your cortex to accurately interpret the reasons for your reactions and feelings is an important tool in coping with anxiety. Keep your focus on what you know about the language of the amygdala to prevent your cortex from increasing your anxiety by offering additional reasons for amygdala-based fear. Thoughts from your cortex about your physical and emotional reactions may not bear any relationship to why the amygdala is responding. Your amygdala may be reacting to a dog because when you were five your grandmother's dog jumped on you and knocked you over, and the amygdala has seen dogs as a threat ever since. The cortex may not even recall this incident. The amygdala is not specifically thinking, *This dog is going to bite.* That thought comes from the cortex, which is trying to make sense of the feeling of fear being associated with this particular dog. In the same way, images of the dog's teeth sinking into your arm come from the cortex, not the amygdala. (And those images are not very helpful!) Luckily, your cortex can learn to correctly interpret the reaction you are having by saying, "My amygdala is being triggered by this dog, but I don't know whether this dog is dangerous." You're having very real reactions and feelings, and you know they come from the amygdala, but you aren't assuming they're justified. That might not be something you say to a friend, but it helps *you* know what is happening, and it doesn't produce erroneous interpretations that cause additional amygdala activation. (We acknowledge this doesn't tell you what to do in the situation; we'll explain that step in the next chapter.)

Once you understand that the amygdala is not logical and that, even though your emotional and physical reactions are very real, you cannot always trust those reactions, you're ready to have a very different relationship with the amygdala. Here's how: The amygdala makes sense in its own way, and it means to protect you, but you want to be the one in control and not let the amygdala take over your life. Using the language of the amygdala, you can better interpret where your difficulties are coming from and teach both the amygdala and the cortex to respond differently so that you have more control over your life. Well-meaning family and friends may not understand what you are dealing with, and they may think that logical arguments and reassurance should be enough to help you overcome your OCD. But you know better, because the amygdala does not learn that way. It's important that *you* understand your amygdala and learn how to manage it, even if others don't understand.

A Matter of Timing

One final important aspect of the language of the amygdala is the *timing* of the amygdala's production of the defense response and the accompanying fear and anxiety. Let's take the example of being afraid of dogs. The amygdala will produce physical and emotional reactions at any sight or sound of a dog, typically *before* the person is in direct contact with the dog. This timing makes sense. The most effective FFF response would need to occur *before* the danger was upon the person. It makes no sense to have the FFF response occur when a tiger is jumping on top of you; it's much more helpful if it occurs earlier, at a time that allows escape or preparation for fighting. This could be why fear can occur even when you just see a dog across the street or begin thinking about visiting a person who owns a dog.

As a result of their connection to the defense response, anxiety and fear are typically anticipatory, and they peak *before* the threatening object or situation is encountered. We see this frequently in our lives and should take note of it because the knowledge can be helpful. It lets us know that the worst fear does not come when we expect it—it actually comes *before* we are in the situation. Fear increases as we approach a situation, peaking just before the situation occurs, and then decreases once the situation is encountered if danger does not appear. Students tend to be most afraid before taking an exam, but once they start the exam the fear goes down. Often a person will say, "If I'm this afraid before I even pick up the phone to call, how can I possibly get through the phone call?" The truth is, when you are putting off making a phone call, staying in the time period before making the call, you are suffering the most.

Most people have no idea that the worst fear occurs *before* the event, and the amygdala will typically stop producing the defense response if danger does not appear in the situation. (That is assuming the cortex does not make the situation worse than it is. More about that in later chapters.) When you're in the process of coping with a frightening situation, knowing about the timing of anxiety can help you recognize that pushing through your anxiety and facing the situation typically makes anxiety go down, not up. For example, Josephina often thought she had a good marketing idea during company brainstorming sessions, but she always became so anxious just before speaking that she kept her idea to herself. She mistakenly assumed that this anticipatory anxiety meant her ideas were not good enough to share. Her anxiety made her stop and doubt her idea. When she pushed through her anxiety and spoke up, she got such positive responses that she realized her anxiety did not mean her ideas were not promising. She also realized that as soon as she started explaining her idea, her anxiety decreased. Josephina's anxiety had been keeping her from showing her strengths. Just another reminder that the amygdala does not respond in logical ways!

In summary, the amygdala communicates through bodily reactions and emotional responses, and the thoughts that we have about anxiety and fears are not part of the amygdala's language, but come from the cortex's interpretations. The amygdala is not logical; it learns on the basis of pairings, not reason. The amygdala is capable of responding before the cortex does, which can be very useful at times, but it is also capable of responding in error when no danger exists. The timing of the defense response and anxiety follows the pattern of typically occurring *before* a threatening situation, peaking just before the situation is encountered, and decreasing once you are in the situation as long as no danger appears. Keep in mind that because your thoughts about your anxiety come from the cortex, they may not bear any relationship to why the amygdala is responding. If you want to understand your amygdala, don't listen to the cortex. Remember the language of the amygdala, especially the fact that it learns based on pairings. If you help your cortex learn the language of the amygdala, it can help you change your relationship with the amygdala in some very useful ways.

Let's consider an example to illustrate the association-based language of the amygdala. When Monica's home was in disarray, she often became anxious. If dirty dishes were piled in the sink, or if clothing and toys were scattered around the living room, she became somewhat panicky. Just the sight of the disorder led to anxious feelings. If she thought about friends coming over, or even a neighbor dropping by, her panic increased. She did not want anyone to see her home looking this way. She knew from her feelings of nausea and dread that her amygdala was involved in creating this anxiety, and she wanted to know how her amygdala had become this way. Was her amygdala prewired to respond to disorder in this way? Or had it learned to react this way?

When Monica's therapist explained that her amygdala had most likely learned to create this fear through pairings and that messy rooms must have been paired with some negative experience, Monica was able to understand why her amygdala reacted this way. Monica explained that her parents had frequently fought whenever their home was disordered, and that her father had told her mother that he wouldn't take her out until she cleaned up the house. This childhood experience likely caused the amygdala to produce fear in response to messy rooms, because messy rooms had been associated with arguments, criticism, and threats. A simple diagram can help illustrate the way in which Monica's amygdala turned a situation like a messy room into a trigger for anxiety. (See figure 9.) A messy room was paired with a negative event, her parents arguing, and the amygdala came to respond to a messy room as an indication of danger. You can apply this diagram to various situations to help you look for the language of your amygdala in your own experiences.

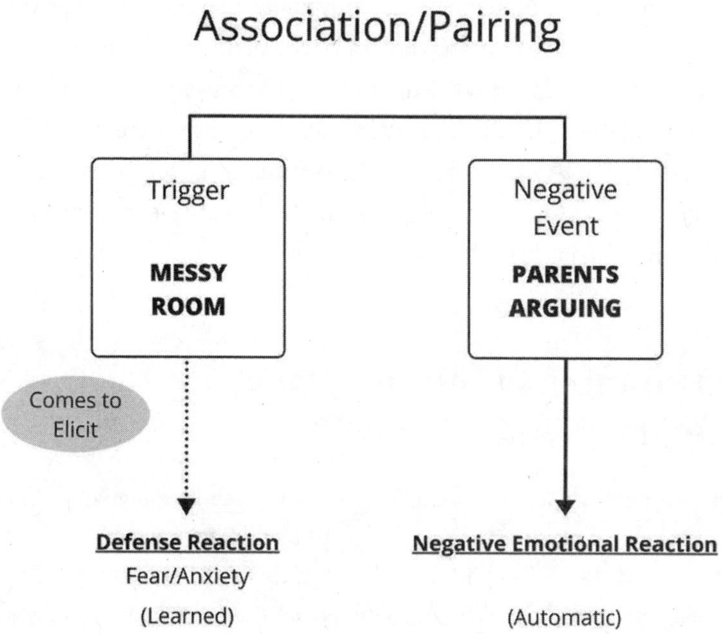

Association/Pairing

```
                    ┌──────────────────────────┐
                    │                          │
         ┌──────────┴──────────┐   ┌───────────┴─────────┐
         │      Trigger        │   │    Negative         │
         │                     │   │      Event          │
         │     MESSY           │   │    PARENTS          │
         │     ROOM            │   │    ARGUING          │
         └─────────────────────┘   └─────────────────────┘

  ╭────────╮  ┊                              │
  │Comes to│  ┊                              │
  │ Elicit │  ┊                              │
  ╰────────╯  ▼                              ▼
     Defense Reaction          Negative Emotional Reaction
        Fear/Anxiety
         (Learned)                     (Automatic)
```

Figure 9

Whenever a situation or an object is associated with an experience that causes a negative emotional reaction, the amygdala creates a memory that turns that situation or object into a trigger for anxiety. Whenever that memory is activated, in this example by seeing or even thinking about a messy room, the amygdala will produce a defense response. Based on the way the amygdala learns, any situation or object that's associated with a negative experience can become a trigger for anxiety. John finds that turning right in an automobile becomes a trigger for anxiety after it is paired with having an accident while making a right turn. A birthday cake came to be a trigger for anxiety for little Emma after it was associated with children making fun of her for not being able to blow out her candles.

Luckily, however, the amygdala is capable of new learning, and people like Monica, John, and Emma can learn how to teach the amygdala to stop responding in this way.

As your protector, the amygdala is constantly on the lookout for anything that may help it predict danger. If any situation or object is associated with a bad experience, the amygdala marks that memory as emotionally relevant to you. In your brain, the memory of a messy room or the memory of a birthday cake is modified by the amygdala to be identified as dangerous. This is how a trigger for a fear response is created. What is paired leads to "SCARED!" The object or situation itself doesn't have to be threatening—as long as it was paired with a negative experience, it can be turned into a trigger for fear and anxiety.

Diagramming in the Language of the Amygdala

Consider some examples to practice interpreting the language of the amygdala. Create a diagram like the one in figures 8 and 9 for each example below. Identify the trigger and the negative event in these three examples. In most cases, these are the only parts of the diagram that you need to figure out. The trigger is typically something that most people consider neutral, and the event is something that most people would consider unpleasant. It is only after the pairing of the two that a person comes to respond to the trigger with anxiety. (Correct answers can be found at the end of the chapter.)

1. Your mother gives you an exasperated look before she yells at you, and the amygdala tags that exasperated look as dangerous, so the next time you see that exasperated look on her face, it is a trigger for anxiety.

2. A woman is assaulted by a man wearing a particular brand of cologne, and after that experience, the smell of that cologne provokes fear.

3. A child asks a teacher for assistance and gets a critical and embarrassing response. For months after this experience, even the thought of approaching a teacher to ask a question is frightening to the child.

What Starts in the Amygdala Does Not Always Stay There!

We will return to discussing Monica's anxiety about messiness in her home to illustrate how a fear that starts in the amygdala can often affect how the cortex responds. We discovered that Monica's fear of messy rooms came from her amygdala. But, if we look closely, we can also see that Monica's cortex became involved in interpreting her amygdala-based fear of messy rooms. When she felt anxious about a mess in her home, Monica's cortex looked for something for her to be afraid of, and it came up with the concern that if other people saw her messiness, they would think poorly of her. The cortex could imagine critical responses that her friends might make about her. This analysis of the situation by the cortex is *not* why the amygdala is responding the way it is. Monica's amygdala had learned on the basis of the pairing of messy rooms with the distressing experience of her parents arguing, and her fear developed when she was a child, before she was even responsible for her own home. But as Monica imagines what other people might say about her if they see her home, the amygdala becomes *further* activated by cortex-based thoughts and images of other people reacting negatively to her having a messy house. She can begin obsessing about this and develop all kinds of ideas about how she compares to others, judging the appearance of each room, feeling overwhelmed by what she wants to straighten in the house, and so on. As you can see, the cortex can take an amygdala-based fear and run with it! What starts in the amygdala does not always stay in the amygdala. This is why we will devote part 3 to explaining how to manage the cortex. But before doing so, we'll teach you how to use the language of the amygdala in ways that can calm the amygdala and, just as importantly, teach it to respond differently to triggers.

Answers for Diagrams

1. Trigger = Exasperated look   Negative Event = Yelling at you
2. Trigger = Cologne   Negative Event = Assault
3. Trigger = Asking a question   Negative Event = Criticism

CHAPTER 9

Teaching the Amygdala

Now that you've learned the language of the amygdala, it's time to focus on how you can teach it to respond differently. Teaching the amygdala is key in changing your anxiety and how it fuels your obsessions and compulsions. You must *show* the amygdala that you're not in danger in order for it to make new memories in your brain that will identify triggers as safe, not as dangerous. In this chapter, we'll guide you through the process of retraining your amygdala to react differently to triggers and help you learn to resist obsessions and compulsions, which only serve to strengthen the amygdala's responding to the trigger as if it is dangerous.

Based on the research evidence, the most effective psychological treatment for OCD is cognitive behavioral therapy with *exposure with response prevention* (Koran and Simpson 2013). In fact, evidence shows that people with OCD who have not been helped by medications can benefit from this approach, often called ERP (McLean et al. 2015). ERP exposes the amygdala to situations or objects that are triggers and gives it the experiences it needs to learn to respond differently—which you will recognize as the language of the amygdala. ERP requires a person to tolerate the defense response (and anxiety) during the process of exposure to triggers and to resist using any compulsions, an experience that is challenging for anyone with OCD. If other ways of teaching the amygdala were effective, we would recommend them, but, in order to change how the amygdala responds to a trigger, it needs to *experience* that trigger, whether the trigger is feeling something sticky on your hands, giving a presentation in front of your coworkers, or having a thought about your partner cheating on you. Nothing that we say to the amygdala, and no amount of reassurance, will prompt it to modify neural memories in the way that direct experience does. Direct experience with the trigger is the best approach for prompting the amygdala to make new memories (LeDoux 2015).

Here is the painful yet powerful truth that will allow you to overcome the limits of your OCD: when the amygdala experiences a trigger, the memory structures about the trigger are activated and so is the amygdala (LeDoux 2015). The amygdala cannot experience a trigger without producing the defense response, and to teach the amygdala to make new memories, *we need to give it experience with the trigger.* This gives you the opportunity to teach the amygdala something new about this specific trigger. Learning in the amygdala will not occur unless you activate amygdala neurons to make new connections between neurons. Remember this phrase: *Activate the amygdala in order to generate new connections.*

Communicating with the Amygdala

You've come to understand that the amygdala learns through association and does not change memories based on logic, reasoning, or reassurance. Those arguments are the realm of the cortex. One of the most frustrating situations for therapists attempting to treat OCD is the knowledge that nothing we say to our clients has much influence on the amygdala. Even reading this book is not changing your amygdala in any way. The information you learn in this book impacts your cortex, and you can use the information to make plans to influence your amygdala in various ways, but the amygdala is not changed directly. Your amygdala learned to produce anxiety on the basis of *experience*, and in order to learn something new, your amygdala will need to *experience* something.

You've already learned that some specific experiences you can provide for your amygdala will change its level of activation and help calm it: deep, slow breathing, especially when combined with muscle relaxation; physical exercise, especially a regular exercise program; and lengthy periods of sleep with plenty of REM sleep. But, if you want to change the memories that determine the way the amygdala responds to a trigger, you need to focus on communicating with the amygdala in the language of pairings, because that's how it learns to respond to the world. Neurons that fire together wire together, and we need to target this wiring process if we want to teach the amygdala something new. We can get the amygdala to create new connections if we focus on presenting *experiences* in which the trigger is *not* paired with a negative event. In other words, you need to expose yourself to the trigger in a situation in which nothing negative occurs. The amygdala will learn, after some repetition of this experience, that the trigger is *not* associated with danger, and it will create new memories and stop producing the defense response in response to the trigger. This is how you can stop the amygdala from creating the defense response (and resulting anxiety) that enables OCD to dominate your life.

Exposure

Exposure is an essential part of ERP. *Exposure* involves repeatedly coming into the presence of a trigger and making sure that nothing negative occurs. This could mean attending a potluck and eating food prepared by others without anything negative occurring. Or deliberately having thoughts about your mother dying without any harm befalling her. We should note that one negative thing does occur during exposure: you will experience the defense response—and anxiety. Because exposure causes the amygdala to produce the defense response, exposure is distressing and anxiety provoking, and this is the reason that many people avoid this type of experience, even though *it is exactly what the amygdala needs* to learn to respond differently. When a person experiences exposure repeatedly, activation in the amygdala decreases as new memories are formed (Roy et al. 2014). If exposure were easy, more people would be retraining the amygdala without any coaching or assistance from therapists, but most people avoid exposure to what they fear, so the amygdala continues to produce the defense response for years and years without ever learning anything new about the triggers that drive OCD behaviors.

Take the example of a boy who was knocked down by a dog when he was four years old. Each time he sees a dog, he feels anxiety and does everything he can to get away from the dog. He could get to adulthood without ever teaching his amygdala that not all dogs are dangerous. Without exposure, the amygdala will continue to operate as if all dogs indicate danger, even though that is not a correct assumption. If his parents decide they want to help him get over his fear of dogs, they need to have him experience dogs repeatedly without anything negative happening. As you see in figure 10, if he repeatedly spends time around a dog, and the experience is either neutral or positive, rather than negative, a new connection will be formed by the amygdala, and, instead of anxiety, a calm response will occur when he sees a dog. The exposure to dogs can occur gradually, with the first experiences being farther away from the dog, and later experiences involving getting closer and closer to the dog, and ultimately touching the dog. Exposure also works when it is done more quickly, but most individuals prefer a gradual approach.

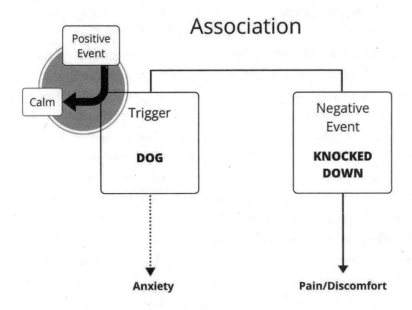

Figure 10

Keep figure 10 in mind to remember how to teach the amygdala in the language of pairing, which it understands, so it can learn to respond differently. The amygdala learns to make new pairings from new information being presented. This can occur through observation too. This is why it can help the amygdala to see another person carrying out the tasks associated with exposure (Olsson, Nearing, and Phelps 2007), whether those tasks involve touching a dog or eating candy from a candy jar in the therapist's office.

Because it is challenging, exposure should be carefully targeted to assist you in meeting goals in your life. It's not necessary to teach the amygdala new responses to all the triggers you can identify in your life. But if certain triggers keep you from getting to work or going to bed on time, discourage you from attending events that you'd like to participate in, or prevent you from advancing at your job, that is a problem. If a trigger is interfering with a goal that's important to you, you need to teach the amygdala to respond differently to it.

Consider Lupita, whose goal was to get past her constant doubts about her effectiveness as a CEO. She was obsessed with the belief that she had fooled everyone and was not worthy of her position. These thoughts were often the first thing she thought of in the morning, and the thoughts she had to get out of her head before she could fall asleep. Lupita was the CEO of a company, effectively supervising many others, but found herself feeling very anxious at many professional conventions and company events. She thought these feelings meant that she did not have the authority to be a CEO, and she was consumed with constant doubts about whether she deserved her position.

When Lupita spoke to a therapist, she discovered that her doubts centered around her difficulties being with loud, boisterous men whom she supervised, and how she always left company parties and conference meetings feeling intimidated and incompetent. When she learned the language of the amygdala and was asked to consider why her amygdala was reacting to these loud male employees as a trigger, she realized it was connected to having an alcoholic father and memories of his loud, aggressive behavior in her childhood. Her eyes opened wide and she asked, "Is my amygdala still five years old?" In a way, her amygdala was acting like her five-year-old self because it had never learned that it could be safe for her to be around loud men. She could logically say that she wasn't in danger because she was the boss, but her amygdala had not learned this lesson. In order to stop feeling anxiety and intimidation, she needed to use exposure and stay in the presence of these loud men and not flee from the situation, until her amygdala learned that the situation was a safe one.

So ERP should start with selecting goals that are important to you and identifying any triggers that set your amygdala off in those situations. You may want to attend family events and holidays, but your young nieces and nephews may be a trigger for you. You may want to finish writing your report for the committee, but every sentence you write that is not "exactly right" causes too much anxiety. Once you select a goal and identify the trigger or triggers that block you from that goal, you can use ERP to help you reach that goal by changing your amygdala's response to the trigger(s).

Identifying Goals and Triggers

Follow these steps to identify what goals you have that are impacted by your OCD and what triggers interfere with accomplishing those goals:

1. Consider different dimensions of your life, including relationships, work, and leisure; for each dimension, consider whether you have goals you want to achieve.

2. Write down each goal, describing what you would like to do (e.g., I want to finish writing my committee reports on time).

3. For each goal, list the triggers that often interfere with your completing the goal (e.g., sentences that don't seem correct).

Were you able to identify important goals, as well as triggers caused by your amygdala that may be interfering with your ability to achieve those goals? If you are willing to be exposed to the triggers through the process of ERP (ideally with a trained therapist), you will teach your amygdala to make new connections. Therapists can help you target the triggers that are having a negative impact on your life, so that you can teach your amygdala to respond differently, allowing you to accomplish your goals.

A note about identifying triggers: Identifying triggers is very important in planning ERP. Discovering the situations, objects, or sensations that activate the amygdala is essential. But you don't need to know how something became a trigger in the first place. Exposure will work to reduce the amygdala's response to a trigger even if you aren't sure how the amygdala acquired the fear, as long as you identify the correct trigger and expose the amygdala to the trigger without any negative events occurring. But you do need to be specific about exactly which part of the situation is triggering the amygdala. In the example of Lupita having difficulties when she was at company events, she needed to know that it wasn't the events themselves or meetings in general that led to her anxiety; her trigger was men being loud. You need to identify specifically what you are triggered by.

ERP is all about staying in the presence of the trigger, which requires tolerance of the defense response, including the resulting emotions, typically fear and anxiety. This can be very challenging. But when you have a better understanding of how the amygdala operates, the nature of the defense response, and the purpose of fear and anxiety, tolerating distress is somewhat easier. Knowing what is occurring in the body, and that it may be uncomfortable but not dangerous, can help. Remember that the amygdala can be completely wrong about whether something is dangerous. But the most encouragement comes from the knowledge that when you tolerate anxiety and fear reactions, you are giving the amygdala a chance to learn to respond differently. We like to joke a bit to help our clients tolerate the amygdala's reaction and their own emotional response during exposure by saying, "Now you have the amygdala's attention! This is just what you want." This is a very different approach than you have probably ever attempted. Rather than focusing on reducing anxiety, you are tolerating it in order to teach the amygdala something very important: *This trigger is not dangerous!*

A note about obsessions and worries: Showing your amygdala that the trigger is not dangerous can be difficult if you are struggling with obsessions and worries that impact you during exposure. For example, Lupita had obsessions and worries that could impact the exposure process. When she was triggered in company meetings, Lupita would experience the defense response and feel anxious. Then she'd think about what her anxious feelings could mean and begin obsessing over doubts about whether she was competent enough to be the CEO. These *thoughts* do not come from Lupita's amygdala; they're coming from her cortex trying to make sense of her amygdala-based anxiety in the meetings. When you allow your cortex to focus on explaining amygdala-based reactions, it searches for thoughts that can justify the feelings of anxiety, and this fuels obsessive thinking and worries in the cortex that have nothing to do with the amygdala's real reason for being activated. In Lupita's case, her amygdala is reacting because of emotional memories related to having an alcoholic father who was loud and aggressive. Ideas about her competency are coming from her cortex and have nothing to do with her amygdala's reaction. This is why it is so important to identify your triggers and interpret them according to the language of the amygdala. Obsessive thinking and worries form as Lupita's cortex reacts to her anxiety by dwelling on doubts about her competency and worrying that others will see her as a fraud. This is how amygdala-based anxiety can fuel obsessive thinking.

Lupita's obsessive thinking is also exposing the amygdala to even more negative experiences at company events and meetings. The company meeting is being paired with negative, frightening thoughts in which she worries she is incompetent. In order for Lupita's amygdala to learn that these meetings are safe, not dangerous, it needs to be exposed to the trigger (meetings with loud men) *without* the meeting being associated with negative experiences, including any negative, anxiety-provoking thoughts. (Obviously, these troublesome thoughts from the cortex are essential to address, but this chapter is focused on teaching the amygdala. We'll discuss ways of changing these thoughts in part 3.)

Response Prevention

The "RP" in ERP stands for response prevention. *Response prevention* means that during exposure, a person is asked not to engage in certain responses. When you have OCD, you may have learned to engage in certain behaviors in order to reduce your anxiety. These behaviors, called compulsions, can become very strong, getting their strength from the way that they provide a quick but temporary reduction in your anxiety. As discussed in chapter 1, *compulsions* are repetitive behaviors or mental acts that a person engages in to respond to a dreaded situation or to reduce distress. They can take a variety of forms, including checking, counting, cleaning, mentally rehearsing, and seeking reassurance. During the exposure process, it's very common to want to use some type of compulsion to cope with being exposed to the trigger. Compulsive behaviors are under your control, although they may be very difficult to resist.

In ERP, you need to agree not to use compulsive behaviors during exposure. The reason for this is that as long as you're allowed to engage in compulsive behaviors to reduce anxiety, the amygdala will not learn anything new about the trigger. Only if the amygdala experiences the trigger without any relief will it learn that the trigger is not dangerous. By obtaining relief from compulsions, you miss the opportunity to teach the amygdala that the trigger is *not* dangerous, and the amygdala will continue producing a defense response to the trigger. The amygdala will learn a trigger is safe only when you stay in the presence of the trigger without engaging in any behavior intended to protect you. Your willingness to tolerate anxiety and distress is very important to accomplish this.

Let's return to the example in which we learned Monica was experiencing any mess as a trigger for anxiety due to messes being paired with her parents' arguments during her childhood. Once you understand how exposure to triggers can teach the amygdala that a trigger is not dangerous, you might ask why Monica's amygdala has not gotten over the fear of messes. Hasn't Monica ever experienced a mess without her parents' arguments or another negative experience being paired with the mess? The answer is perhaps not, especially if Monica has been working to avoid messes all these years. If she feels anxious when her bedroom is disordered and dirty, and she escapes from the mess by avoiding it or by cleaning it, she feels relief. Unfortunately, each time Monica escapes from a mess, even by cleaning it, she is maintaining the anxiety. She has not taught the amygdala anything new about messes.

You may be surprised to learn that the use of cleaning to cope with her anxiety about messes has *strengthened* Monica's anxiety about messes.

Her amygdala has not only learned that *Mess* is associated with danger; it has also learned that *Cleaning* is associated with relief. In fact, the compulsive behavior of cleaning produces relief that seems to validate the dangerousness of the mess. Monica, like most people, does not know how to teach the amygdala that a mess is not a danger. When she cleans to remove the mess, the amygdala cannot experience exposure to the mess long enough to have a chance to learn anything new. While cleaning *temporarily* reduces Monica's anxiety, it actually is keeping her amygdala from learning to respond differently.

Compulsions develop because they reduce anxiety temporarily, which feels very good. Unfortunately, compulsions also prevent effective exposure and ensure that the amygdala will continue to produce anxiety in response to triggers. One point that may not be obvious is that the compulsion does not have to be related to the trigger for it to produce relief. In Monica's case, her compulsion is cleaning, which removes the mess, so it is clearly related to the trigger. But a compulsion can also be a behavior that has nothing to do with the trigger. Timothy's compulsion was saying license plate numbers out loud, which did not seem to have any apparent relationship to the trigger of driving in congested traffic, but it served to reduce his anxiety. As she went to bed each night, Shanice had to check under her bed and in her closets for an intruder whenever she had thoughts about her parents dying, and this relieved her anxiety. The compulsion does not need to be related to the trigger. Anything that has succeeded in helping the person find relief from anxiety will be repeated again and again to obtain relief, whether it is logical or not.

Now you can see why the RP part of ERP is so important. As long as you're allowed to engage in compulsive behaviors to reduce anxiety, the amygdala will not learn anything. You have to activate the amygdala for a sufficient amount of time for it to generate new connections. This is why the most effective exposure for OCD will include response prevention, requiring that you refrain from engaging in compulsive behaviors such as escaping, cleaning, seeking reassurance, checking, and so on, during exposure.

ERP in Action

Let's examine how ERP helped Monica with her cleaning compulsion. Monica had lived for years with the process of temporarily reducing her anxiety about messes by cleaning compulsively, without addressing the real problem of her OCD. Finally, something happened that made Monica decide she needed to seek some help. She had just had her third child and was finding that her house was messier than it had ever been before. She reported her anxiety was going through the roof. She couldn't keep up with the toys and laundry that were all over the house and the dishes that were piling up in the sink. She would be up in the middle of the night feeding her infant, and then, seeing the house in disarray, would be up for hours trying to clean her home. When Monica came in for therapy, she said she had never been so anxious, self-critical, and discouraged in her life. She didn't know how she could keep living with the situation. She wanted to set the goal of eliminating anxiety from her life, but the therapist knew that goal was impossible. They agreed on the goal of Monica finding a way to enjoy being a mom again, especially being able to feel comfortable focusing her time and attention on her children.

In therapy, Monica learned that her amygdala was the source of her anxiety, and, as noted in chapter 8, she recognized how her amygdala had learned to react to a mess with anxiety as a result of a messy house being paired with her parents arguing. In order for Monica to enjoy being a mom and focus her attention on her children, her amygdala needed to learn to produce less anxiety in response to the messy house, and ERP was recommended. The amygdala needed to learn that a mess did not pose a danger. This meant exposing Monica's amygdala to a messy house and having her agree not to clean, without anything negative occurring.

Exposure therapy involved telling Monica that for two weeks she was to focus on attending to her children and herself and refrain from housekeeping. She was encouraged to allow disorder to happen in her home and to accept it as a way to teach her amygdala it was not a danger. She worked with the therapist to identify what housekeeping tasks were essential to keeping her family cared for and healthy, and which tasks she could deliberately neglect. She was allowed to wash dishes during the day, in order to keep enough dishes clean to feed herself and the family, and to care for food appropriately by keeping it refrigerated and stored safely, but was limited to doing laundry on the weekend when her husband was available to assist her. She agreed to a designated list of *prohibited behaviors*, including vacuuming, dusting, and putting toys and clothes away, and she was also prohibited from doing any chores after 10 p.m. because her sleep was important.

Because the goal was to expose Monica to a messy house without anything negative occurring, her husband needed to understand how to support this process. Before implementing this plan, Monica's husband was invited to talk with Monica and the therapist about the purpose and goals of ERP and to share his view of the housekeeping situation. In contrast to Monica, her husband thought that the house being messy was understandable given his long work hours and Monica's responsibilities for the newborn and two older children. He readily agreed to support the plan. Luckily, he was much more accepting of disorder in the house than was Monica. He wanted Monica to be less anxious about her life, and he assured her that he understood the purpose of exposure and would not be irritated. This was essential, because if he had had a negative response, this would have been paired with the disordered house and defeat the whole purpose of exposure.

During this period of ERP, the therapist instructed Monica to deliberately notice the disorder in her home but focus on caring for her children and herself rather than on house cleaning. The therapist told Monica that she should expect her anxiety to increase but assured her that her anxiety would go down as nothing negative occurred and her amygdala began to learn. When Monica expressed worries about how others might judge or criticize her due to the disorder in her house, her therapist reminded her that she was trying to teach her amygdala that any mess was safe and that associating these kind of thoughts (from her cortex) with the mess in the house were counterproductive. If she engaged in these thoughts, she was making exposure less effective. She laughed when the therapist suggested that, if these thoughts came, she should imagine greeting visitors with a phrase that the therapist's grandmother often said: "If you came to see the house, take a look and leave. If you came to see me, clear off a seat, sit down, and we'll have a cup of tea!"

After one week, Monica said that she had felt very anxious for the first couple of days but had still managed to prevent herself from engaging in tasks that were prohibited. She was most proud of how she had allowed some dishes to go unwashed overnight when she hadn't been able to do them by 10 p.m. She was surprised at how quickly her anxiety had decreased once she pushed through the initial anxiety of going to bed with dishes in the sink, by saying sleep was more important for her to be a good mother and she would do them in the morning. She said that she had focused on deliberately activating the amygdala, and had even told herself, *The amygdala wants me to do dishes, but I'm not going to.* She recognized a clear reduction in her anxiety about messiness in the house after one week focused on exposure. After two weeks, Monica reported that she felt she was managing very well; her house was more disordered than it had ever been, but her anxiety was much improved.

Monica reported that she felt her amygdala had learned an important lesson, and she felt much more in charge of her life. She was cleaning only during the day and focusing on her goal of finding joy in being a mother and putting her children first. She believed she could continue using a plan of no nighttime cleaning for the foreseeable future. She worked with the therapist to decide how to continue to limit her time on specific chores, basing her life not on managing her anxiety, but on her own ideas of being a happy and effective mother. She still had some anxiety about messes at times, but could push through it.

How to Use ERP Successfully

ERP can be a very challenging process, and working with a therapist specifically trained in ERP is strongly recommended because these therapists know how to coach you through the process. ERP therapists know how to provide the correct exposure experience to activate the amygdala at the best level so that it learns that the trigger is not dangerous. Therapists also know how best to provide encouragement and not force a person to move too quickly through the exposure process. You'll be expected to go beyond your comfort zone, but not beyond the safety zone. Sometimes the therapist participates in exposure with you, at times doing the same things that you do, at least in the first exposure sessions. Doing exposure with an experienced therapist has been shown to be especially effective and work more rapidly (Lang and Helbig-Lang 2012). Although the presence of a therapist makes exposure more effective, especially in initial sessions (Gloster et al. 2011), eventually a person should do exposure independently (Craske and Barlow 2007).

ERP works best for reducing a fear that is blocking you from accomplishing a goal that is important to you. This helps you have the motivation you need to push through the anxiety. Also, instead of taking on a situation all at once, you may need to break the exposure process down into steps by describing a series of situations. The process should be carefully planned out with a therapist to identify the right level of anxiety to start with. Usually, we ask clients to rate a variety of situations on a scale of 0 to 100, with 0 being no anxiety at all and 100 being intolerable anxiety. We then put these situations into a list, called an *exposure hierarchy*, with the least anxiety-provoking situation first and the most anxiety-provoking one last.

For example, if Ricardo has difficulty attending church potlucks due to fear of contamination of food prepared or touched by others, going to the potluck would not be the first step. The therapist would ask Ricardo to identify and rate other situations that provoke anxiety about consuming food touched by others. He might come up with various examples like eating peanuts from a can touched by the grocery clerk (rated as a 10) or eating a piece of candy from the candy jar in the therapist's office (rated as a 40), which are less anxiety provoking than eating at the church potluck (rated as a 90). We usually start with situations in the 30 to 40 range, because these activate the amygdala enough that a person can clearly experience the decrease in anxiety that occurs after the amygdala has been exposed to the trigger for a period of time with nothing negative occurring.

Learning to tolerate some anxiety is essential in exposure. For each ERP session, the key is to maintain exposure to the trigger without performing any compulsions. Often the therapist is present at exposure sessions and is also doing whatever you are expected to do. The therapist provides support and encouragement and reminds you that if you wait, your anxiety will go down naturally. As you put yourself into these exposure situations, you can actually *feel* your amygdala learning. A change occurs in your level of anxiety that allows you to experience the amygdala beginning to respond to the trigger in a new way. This is very empowering!

Note that as you practice ERP, the cortex can definitely interfere with exposure. If you're focused on thoughts that activate the amygdala, you are working against a central goal of exposure, which is to show the amygdala that the situation is safe. For example, if you're giving a speech to a group, you're trying to experience the situation *without anything negative occurring.* If you keep looking at a person who is frowning (perhaps puzzled or concentrating) and thinking to yourself, *He doesn't think I make sense. I must sound stupid,* you're likely to be maintaining amygdala activation by pairing the situation with a negative event instead of a neutral or positive one.

Helpful Guidelines for ERP Sessions

1. Rate your anxiety on a scale of 1 to 100 at the beginning of the session and at regular intervals during the session. Saying a number out loud is sufficient.

2. Do not engage in any compulsions or behaviors designed to reduce anxiety.

3. It is okay to engage in deep breathing to promote calming the amygdala to a manageable level, if you wish, but anxiety will still occur. The goal is to activate the amygdala at a level you can tolerate and remain in the situation.

4. You may complain about the anxiety and other symptoms that you experience; just don't leave the situation.

5. Remind yourself that your anxiety and other bodily responses are an indication that you are putting the amygdala in the best situation to learn.

6. Try to avoid focusing on negative thoughts or interpretations about the situation during exposure because these can further activate the amygdala.

7. Stay in the exposure situation without performing any compulsion until your anxiety has decreased by at least half from the initial rating. Anxiety does not have to go down to zero, but it should decrease noticeably.

8. Repeat exposure to the same situation multiple times in order to teach the amygdala most effectively. Stick with the situation until the level of anxiety at the beginning of exposure is a manageable level for you.

9. Your exposure experience should be tailored to your own individual needs; the number of sessions needed will vary depending on the person and the exposure situation.

One of the most empowering experiences you can have as a person dealing with OCD is learning that you have some control over the amygdala and can teach it to respond differently. The truth is, the amygdala is very capable of learning from new experiences, and it is often we ourselves who prevent it from having the experiences that it needs to learn. If you keep your world restricted to make sure that the amygdala is activated as little as possible, by avoiding triggers or always using compulsions to relieve anxiety, you're keeping the amygdala from learning and changing. This is natural, because we all have a tendency to avoid anxiety-provoking situations. But the amygdala needs those situations to learn. If you have a situation in which you want to be in control instead of the amygdala, ERP provides you with the ability to teach the amygdala to respond differently.

A Final Note About Compulsions

In this chapter, you have learned how to teach the amygdala using the power of exposure. But hopefully, you have also developed a new perspective on compulsions, which are often mistakenly seen as useful coping strategies. Compulsions can give you a quick fix for your anxiety. When you carry out the compulsion, anxiety is reduced, the relief that you feel is rewarding, and that makes the compulsion stronger. Pretty soon a vicious cycle forms in which the compulsive behavior is the way you begin to manage the anxiety. Feel anxiety? Perform the compulsive behavior. Anxiety decreases. Relief. What's the problem? The problem is that the decrease in anxiety is only temporary. The amygdala has not learned anything, and it keeps producing anxiety day after day. The cycle will continue, providing temporary relief, but anxiety will also continue. The goal of ERP is to stop the reliance on compulsions in order to directly change the amygdala, *the source of anxiety that is fueling the compulsions*. We want to teach you how to have a *lasting* effect on anxiety, not a temporary one, by focusing on its source. ERP is not easy, but it can produce lasting and life-altering effects!

In summary, in part 2 you learned about the nature of the defense response created by the amygdala and how it results in the anxiety that fuels your obsessions and compulsions. You learned that to get to the root of OCD, you need to understand the role of the amygdala. You learned various ways to calm the amygdala, and perhaps most importantly, to teach the amygdala to respond differently. This gives you the ability to take control back from the amygdala. In part 3, we will address ways to help you keep your cortex from getting caught up in the debilitating patterns that maintain OCD.

PART 3

Managing Your Cortex

CHAPTER 10

How to Manage Your Obsessive Thinking

Whether your obsessions involve thoughts, images, or urges, the problem is that obsessions keep coming back. They intrude into your life and take up precious time that you could be using in better ways. They also produce anxiety that not only causes you distress, but may also cause you to engage in compulsions or avoid certain situations. You want to be in control of your obsessions, but they just don't seem under your control. Our aim in this chapter is for you to understand the best way to manage your obsessions. To do this, you need to put together all that you've learned about what is happening in your cortex and how it affects your amygdala.

You now know that the thoughts, images, or urges that make up your obsessions come from your cortex and that they are stored in neural circuitry. You've learned that cortex circuitry follows the law of "survival of the busiest" (Schwartz and Begley 2003, 17). The circuitry that you use the most becomes the strongest, and the circuitry that you don't use becomes weaker. This means that the more you think about or discuss your obsessions, the stronger they become. But it's important to understand that when we're advising you against using the circuitry, we are *not* saying that the thoughts, ideas, or urges are dangerous. Your obsessions themselves are not a danger. We therapists are not afraid of your obsessions; many of us obsess ourselves. We know that obsessions are troubling, but they are *only thoughts*, and we need to start with this important point: "Don't treat your thoughts as if they are dangerous!"

To help your amygdala learn that your obsessions are not dangerous, sometimes you need to face the obsession and discuss it in detail with someone. This emotional processing (Foa, Huppert, and Cahill 2006), similar to exposure, helps your amygdala learn that these thoughts are not dangerous. This shows the cortex that you can safely think about the topic (of your father dying, or your classmates ridiculing you) and it doesn't mean that these things will happen or that you can't handle them if they do. Although it will be stressful to consider these kinds of thoughts—or even dwell on them until your anxiety goes down—doing so will not make anything happen in the real world. If thinking about something made it happen, we could spend time thinking about winning the lottery and end up being millionaires! Thoughts by themselves can't cause things to happen. Going through this kind of experience of discussing your obsession to emotionally process it with someone who recognizes what it truly is—just thoughts—can help you change how your cortex and your amygdala respond to your obsessions.

Defusing Obsessions

To defuse an obsession is to stop treating a mere thought (whether it is an idea, image, or urge) as if it reflects reality. The truth is that many thoughts that go through our heads are simply noise. Remember that not only people with OCD but just about everyone (80% or 90% in most studies) has troubling intrusive thoughts (Radomsky et al. 2014). Odd thoughts about killing oneself or harming someone can regularly pop into people's minds. Most people let those thoughts go. They say, *That was a weird thought* and move on with their day. The mistake that people with OCD make is dwelling on their thoughts and giving them more attention. "Your distressing thoughts don't deserve all this attention," we therapists say. Too much focus on an intrusive thought will strengthen that thought to become an obsession. To defuse a thought, you need to remind yourself that it's only a thought and not give it the attention that will strengthen it. If you have already focused on it and strengthened it to the level of obsession, then you have some work ahead of you, and this chapter will lead you through it.

In his book titled *Stopping the Noise in Your Head*, Reid Wilson (2016) said, "Don't get caught up in the content!"—meaning that the content of your obsessions does not matter. Whether you're thinking about possible rejection, sexuality issues, deadly diseases, contamination, or potential mistakes, it's always the same response from us cognitive behavioral therapists. Just remember it is only a thought, and the specific content doesn't even matter. The problem is not about being rejected, or your sexuality, or a deadly disease. The problem is that *you are having a thought that is stressing you out*. It doesn't matter what it's about. You are caught up with it and feeling it's necessary to spend your time focusing on it. Obsessions need to be starved of attention. That will weaken them.

When clients diagnosed with OCD come in to our office obsessing about whether they ran over a person on their drive to the office, whether they have a sexual attraction to someone they shouldn't, or whether someone they love is going to die, we respond to all of these obsessions the same way: "That is just a thought you are having." Sometimes we take a piece of paper and write the thought down in a thought bubble and hold it over a client's head. It is a *thought*. Whatever the content, it is a still just a thought. Sometimes writing the thought fifty times can help you see it as simply words.

Therapy groups for people with OCD can help them recognize that it doesn't make sense to get caught up in the content. In one such group, Roger is confused about why Joel can be so stressed out about his feces seeming to be a strange color this morning. "Why would you think that something like that means you have cancer?" But then Joel responds, "Why do *you* think because you heard your car hit something on the road this morning that you killed an animal?" In both cases, they are dealing with thoughts. They're caught up in the content of the thought and taking the thoughts too seriously. Each one has different fears that underlie their thoughts: one is focused on health and one is focused on harming others. But they each are being plagued by an obsession, and the problem is the same: cognitive fusion. They are mistaking a thought for reality.

Trying to keep our clients from being caught up in their obsessions is like trying to get the attention of someone when a spider is crawling around the room, and they don't want to take their eyes off that spider. It's just as hard to convince a person to take their attention away from their obsessions. For example, Hannah kept asking her therapist for reassurance that she did not have a sexually transmitted disease, despite repeated tests that showed she was not infected. Her therapist took a large, scary-looking plastic spider and showed it to Hannah. The therapist asked Hannah to look at the spider and then put it on the floor in the corner of the office. When the therapist sat back down, Hannah looked back at the therapist. "Why aren't you watching the spider?" the therapist asked. "It's not real," Hannah answered. The therapist said, "Neither is the sexually transmitted disease you keep worrying about, so how can I get you to try to take your focus off that?" This helped Hannah see her obsession in a different way. A thought about a sexually transmitted disease is no more a real disease than a plastic spider is a real spider.

Thought-Action Fusion

Thought-action fusion is the tendency to believe that if you have a certain thought, image, or urge, you are likely to act in a way that is consistent with it. For example, when Isaak is supervising his little sister and her friend Natalie as they swim in the pool, he notices that Natalie is a cute little girl and thinks she is likely to be very attractive when she gets older. He instantly thinks that his thoughts about Natalie might mean that he's attracted to her, and starts to worry that he might act on his thoughts and do something inappropriate with Natalie. He feels very anxious and horrified at what he has been thinking. He has made the mistake that his thought implies that he will take a specific action. Knowing what you know about anxiety in the brain, can you identify where Isaak's anxiety came from? When he thought of acting on his thoughts that Natalie was attractive, he felt anxious. Is he a pedophile? Isaak is clearly not a pedophile because he was not sexually excited by his thoughts and considering acting on them; instead, those thoughts distressed him and activated his amygdala. The amygdala was reacting not because the behavior was likely, but because the situation that he imagined seemed a dangerous one.

Cognitive fusion and thought-action fusion are not the only beliefs that make a person vulnerable to obsessions. A whole set of beliefs has been identified as providing the context to promote the development of obsessions (Fergus and Wu 2010). When a random, intrusive thought pops up, these beliefs are likely to lead a person to have more concern about that thought and to attend to it in a way that promotes it from a simple intrusive thought to a troubling obsession. In chapter 11, we will examine these self-defeating beliefs, including perfectionism, catastrophizing, and the need for certainty, and help you learn how to replace them with coping thoughts that make you less vulnerable to obsessions. For more on this topic, also see the bonus chapter "The Healthy Use of Worry" at http://www.newharbinger.com/47186.

You Can't Erase Obsessions: Replace Them!

It would be so nice to be able to suppress a thought that troubles you. Unfortunately, these are the very thoughts that capture your attention. Sometimes it seems that OCD knows just what kind of thoughts to dangle out there in order to get us to pay attention. But this is where our human limitation with focusing our attention can work in our favor in defeating OCD. While our brains can do thousands of different things at once, they can only focus attention on one thing at a time. (This is why you can't text and drive. You can't attend to both your texting and driving at the same time; you can only switch your attention back and forth. And that is dangerous!)

Being able to focus our attention on only one thing at a time is a gift to those with OCD. If you can stop thinking about the obsession, you can weaken it and make it less likely to dominate your thinking. And the way to stop thinking about it is not to try to suppress it or erase it. You need to *replace* it with other thoughts. If you focus your attention on *something else*, your brain is not capable of focusing attention on your obsession.

Our brains are not like televisions that you can turn off. The best you can do is change your focus to another channel. It doesn't matter what else you focus on, as long as it's not related to the obsession. Don't stay on the Obsession Channel and argue with your OCD about the obsession. This is still focusing on the obsession. As long as your OCD can keep you focused on the obsession, OCD is in control. You need to change to a completely different channel. Some possible channels could be playing with the dog, watching the news, paying your bills, planning a trip, or calling a friend. If your television has 500 channels, how many channels do you think your brain has?

Don't fall into the trap that if you think about the obsession in the right way, you can find a way to weaken it. There is no way to weaken circuitry by activating that circuitry. When you think for a long period of time about your obsessions, you are creating deep grooves in your thinking processes that make your thoughts likely to flow down the same pathways over and over. Any time you make your cortex focus on a particular thought, image, or urge, you are making that circuitry stronger. In a similar way, when you ask other people to discuss an obsessive thought (or image or urge) to reassure you, you are also increasing the strength of the circuitry. Seeking reassurance means you have recruited someone else into helping you make the circuitry (and your obsession) stronger. Learning that neurons that fire together wire together gave you a new insight into what maintains obsessions. It should also help you see that if you change what you think about, you can change the circuitry in your brain.

Sometimes a good way to begin weakening the circuitry underlying your obsessions is to take a gradual approach. If you can schedule certain times of the day as "obsession-free periods," you can begin to limit your focus on your obsessions without feeling you have to give them up entirely. This can sometimes feel more possible to accomplish. Often you can already identify certain obsession-free periods that are occurring. For example, Shonda could already say that she was keeping herself from obsessing at work. She explained that at work, she couldn't take the time to do that when she was on the clock. But Shonda knew she was obsessing in the morning and when she came home from work. She agreed to extend her obsession-free period when she came home from work, not allowing herself to obsess until she finished her dinner. Then she extended the period to 8 p.m. After accomplishing this, her therapist asked her if she could extend it to 9 p.m. Being able to take control over her obsessing made Shonda more confident, and she said, "Can I just watch Netflix and not worry about obsessing at night?" getting a grin and a thumbs up from her therapist. You can start by limiting your obsessing in a gradual way, and increase it as you go along.

What If My Thought Feels Dangerous?

Something that is often very troubling about obsessions is the anxious feeling you get when you have them. Of course, if a thought produces anxiety, you will probably take some time to consider that thought. It seems to be dangerous. But now that you know *why* a thought produces anxiety, you can take a different perspective on the thought. The thought activates your amygdala because the amygdala watches whatever channel you are tuned in to in your cortex, and if the content of that channel suggests a potential danger, the amygdala takes that danger seriously, whether or not an actual danger exists. Remember that the amygdala cannot necessarily distinguish between thoughts that are reasonable concerns and random thoughts that are just noise. But your cortex should be able to do that now, especially if you've read each chapter in this book!

You've learned that anxiety results from the amygdala producing the defense response, and that the amygdala can often react in error. The amygdala doesn't have the ability to interpret the meaning of an experience in the way the cortex can. You've learned about how the cortex can produce interpretations or worries that can activate the amygdala to produce anxiety, even when the interpretations have no evidence to support them and the worries never end up coming true. You've learned that the amygdala cannot evaluate whether the cortex is focused on something that should be taken seriously, so it often takes thoughts more seriously than needed. You've learned about anticipatory anxiety and how it tends to peak as a situation approaches. Because you've become more educated about what can cause you to experience anxiety, you can be more skeptical about whether feeling anxiety means that a danger exists.

Anxiety is not a pleasant experience, but it does not reliably indicate danger. Anxiety is only a feeling, and not always an accurate indication of what will occur.

Apply what you know about anxiety to your obsessions. If your obsession results in anxiety, why is that? If you have a thought that your partner is cheating on you, or a doubt about whether you have been exposed to a sexually transmitted disease, and you feel anxiety, does the anxiety mean that you should take your thought or doubt seriously? Knowing how your brain operates to produce anxiety, and knowing how the amygdala responds to the cortex, you need to step back from the thoughts you have and remind yourself that they are just thoughts. This is true *even if* you feel anxiety. Just because the amygdala reacts does not mean the reaction is justified. You may have no evidence that the thoughts are legitimate concerns.

Don't trust your anxiety. Anxiety is a feeling that you get when your defense response is activated by your amygdala. You cannot rely on your *amygdala* to judge whether an obsessive thought is true or false, so you can't rely on your *anxiety* to know when a thought is true or false. You may have uncertainty about whether or not the thought is true or false, but you should not believe that the presence of anxiety tips the scale toward "true." This is because the amygdala frequently doesn't know what is true, and it tends to overestimate danger, just to be on the safe side. Many times, when you learn to live with uncertainty, you have much less difficulty with obsessions.

There's another mistake we can make in dealing with the anxiety that an obsession can cause. Because we value our thinking processes, we sometimes learn to think and think and think about our obsession, until we finally think for long enough or in a certain way that anxiety is reduced. Because anxiety goes down after that period of thinking, you feel better, and it feels like you have managed to control the obsession. You really haven't, though, because you were thinking about it, and activating and strengthening the circuitry. Because of this, after a period of time, the obsession comes back again. Then, you go through that elaborate thinking process again, until your anxiety goes down, but each time you spend all this time thinking about the obsession, you're strengthening it more and more. Further, you may feel you must engage in this thinking, and now you don't just have to deal with your obsession; you now have a mental compulsion. How much better would it be if you could focus your cortex on a different channel and discover that you could make your anxiety go down without keeping the obsession circuitry activated? Then you'd be reducing your anxiety and also weakening the circuitry at the same time.

Perhaps the greatest challenge in attempting to stop obsessing is to have the courage to push through your anxiety without falling into the habit of obsessing or performing a compulsive mental process or behavior. Learning to distinguish between the feeling of anxiety and the actual existence of danger can be helpful. Remember that feeling anxiety does not mean you are in danger. It's often like a false alarm going off. You can learn to tolerate it. Remind yourself of how many times you have worried about something that did not happen.

In recovery from surgery, the doctors often get you up and moving, even though there's pain, and tell you that it's important to push through the pain to recovery. You may ask, "Is it okay to be walking, if it hurts?" and the doctor will reassure you that you need to build the muscle, and it will hurt for a while, but the pain does not mean you're doing damage to the recovering area. In the same way, we recognize that enduring anxiety, even if there's no danger, is a tremendous challenge. But pushing through anxiety does not harm you. In fact, this is true courage. Courage is not the absence of anxiety; it's the ability to respond a certain way *despite* anxiety. Pushing through anxiety and focusing on other things is difficult, but responding despite anxiety is often required to take control back from OCD.

Mindfulness

Mindfulness is a technique that has been incorporated into cognitive behavioral treatments for a variety of disorders, including depression and anxiety-based disorders, like OCD (Hayes, Follette, and Linehan 2004). In mindfulness, you attempt to observe your internal or external experiences in the present moment, with a focus on accepting (rather than changing or judging) the experiences you are having (Orsillo et al. 2004). Mindfulness can involve focusing on something as simple as listening to the sounds in the room, tasting a piece of chocolate, or being aware of the movement of breath in your nostrils. It can also involve focusing on complex experiences, like the various aspects of the feeling of anxiety in response to a specific trigger. Learning mindfulness meditation requires practice, and what you are practicing is simply the art of deliberately focusing on what you *intend* to focus on. You become more aware of aspects of your experience that you've been neglecting, and you learn to tune in to these experiences in a deliberate, curious, and exploratory manner.

An important skill that you develop when you practice mindfulness is an increased ability to select what you focus on and maintain your focus on that. This is similar to the idea of changing the channel in your cortex. Mindfulness helps you develop the ability to change channels more effectively by strengthening your brain's ability to focus on what *you* choose. Research shows that after eight weeks of training in mindfulness, certain parts of the brain—the anterior cingulate cortex and the prefrontal cortex—become thicker, suggesting these areas are strengthened by mindfulness training (Holzel et al. 2011). These areas are involved in focusing attention, processing emotions, and detecting errors, and you may also remember that they have been shown to be involved in OCD. Mindfulness techniques help you focus attention in a way that can weaken the circuitry underlying your obsessions because you learn how to replace one set of thoughts with another. The use of mindfulness can also affect the amygdala; Way et al. (2010) found decreased amygdala activity in the resting state for those who showed skills in mindfulness.

The truth is that the human brain tends to be constantly shifting focus and easily distracted. Selecting a focus and maintaining that focus is not a well-developed skill for many of us. Mindfulness helps us learn this skill. This is a great advantage for combatting obsessions because the essence of managing obsessions is taking control of what you focus on. OCD is always bringing up certain thoughts or worries and demanding, "Think about this! What about this?" If you have trouble with focus, mindfulness is a skill that you really need in the battle with OCD. Who is going to control what you focus on? You or OCD?

Because obsessions so often take us away from the present, asking us to endlessly consider the past or repeatedly worry about the future, mindfulness is an amazingly appropriate antidote. With mindfulness, we can focus on the present moment by tuning in to our senses and current experience. If you think about it, our senses are *always* focused on the here and now, so if we focus on our senses, by seeing all the colors of a sunset, smelling and tasting the coffee, or even hearing the sounds of cars going by, we have suddenly anchored ourselves in the present, and this can be a welcome relief from obsessive thinking. OCD often wants us to focus on the past or the future in a way that activates the amygdala and gets us stuck in repetitive cycles of thought in our cortex. Mindfulness can help put *you* in control, rather than OCD.

One of the best ways to begin practicing mindfulness is to tune in to your senses and focus on experiencing them in detail. Please take the opportunity to explore your ability to focus by downloading the mindfulness exercises at http://www.newharbinger.com/47186. After some practice focusing on your sensory experiences, you can focus on emotional responses and how you experience them in your body. You learn to focus with curiosity on the experience, not to judge it or try to change it. You begin to learn how emotional experiences change on their own; in fact, it seems impossible to hold on to an emotional experience for very long without it fading or shifting. You eventually become able to observe your thoughts in a similar, curious, nonjudgmental way. Mindfulness helps us observe and experience our thoughts as thoughts and not experience them as fused with reality or with actions.

So mindfulness not only helps you focus but also gives you a new way to experience what goes on in your brain. You can improve your tolerance for anxiety and other challenging emotions. You can change how you view the thoughts going through your head. You start to see that an intrusive thought can be observed and not experienced as a threat. Alma found that after she practiced mindfulness, she felt more capable of "standing back and observing my thoughts" instead of being frightened of them. She said she was really helped by something her therapist had said: "Consider your thoughts to be like the cars going by on the busy street here. As long as you stand on the sidewalk and watch the cars, you are safe. When you get all caught up in your obsessions, it's like jumping into the path of the cars." Mindfulness helped Alma observe her obsessions rather than putting herself in the midst of them.

Improving mindfulness skills is much like developing muscles through exercise. As you practice and use mindfulness, you strengthen your ability to focus and maintain your focus more effectively. This increased ability to focus your attention means that you can have more control over how much time you spend on obsessions. But don't expect that your ability to focus will ever become like a firm grip. Attending to something is more like *looking* at it than like *holding* it. That means that it's normal for you to glance away from time to time, but the trick is to bring your attention back to what *you* want to be looking at.

Many people feel that the fact that their attention wanders means that they cannot use mindfulness. But the human brain is a wanderer, and you just need to keep bringing it back with patience and good humor. It's perfectly normal to lose focus and find yourself thinking about something else; just bring your focus back to whatever you intend to focus on. If your focus wanders five times, bring it back five times. Each time you bring it back is like training a muscle; you're strengthening your ability to focus. Try not to see a temporary loss of focus as an error on your part. The error would be to give up trying to focus and just let your brain take control and focus in depth on another topic. As the Buddhists say, "It is okay if a bird lands in the tree; just don't let it make a nest."

We hope you recognize the benefit of mindfulness for people who are trying to manage their obsessions and anxiety. The skills of deliberate focus and nonjudgmental, curiosity-driven observation of one's experience are extremely useful in the process of rewiring the cortex. They can assist you in defusing your obsessions as well as weakening the circuitry that promotes them and replacing it with new circuitry that is less likely to result in anxiety. Being trained in mindfulness is certainly an advantage for those with OCD, and many books, community training opportunities, and online resources are available on the practice.

In this chapter you learned how your cortex is the source of obsessions and that you can change your cortex to be able to resist and weaken them. Knowing that the cortex is governed by "the survival of the busiest," you need to work on replacing certain thought processes with other thoughts in order to rewire your cortex circuitry and weaken obsessions. In the next chapter, we will explore ways to calm the cortex to make it less likely to activate the amygdala.

CHAPTER 11

How to Calm Your Cortex

We have shown how influential your cortex is in activating the amygdala to produce the defense response, resulting in unnecessary anxiety. In this chapter, we will address a variety of ways to calm your cortex in order to prevent this from occurring. When the amygdala is not reacting, and the cortex produces thoughts, images, or emotions that activate the amygdala, we call this *cortex-based anxiety*. Even though the cortex cannot directly produce the defense response, the processes in the cortex can certainly be what prompts the amygdala to react. As you learn ways to combat cortex-based sources of anxiety, you'll have more tools to reduce the anxiety produced by your amygdala. These tools also are helpful in controlling obsessive thoughts.

Cognitive Restructuring

One of the concepts used in cognitive behavioral approaches, adopted from cognitive theorists such as Aaron Beck (1976) and Albert Ellis (Ellis and Doyle 2016), is the idea that some thoughts (or cognitions) are illogical or unhealthy and can create problematic patterns of behavior or emotional responding. Cognitive behavioral therapists focus on identifying and changing thoughts that are self-defeating or dysfunctional, including thoughts that lead to increased levels of anxiety or depression. This process is known as *cognitive restructuring*. Whenever we're trying to change thoughts, we're attempting to modify the cortex in some way. Our thoughts are not just the result of neurological and chemical processes in the brain;

they *are* the neurological and chemical processes in the brain. In cognitive restructuring, you change your thoughts to rewire your brain by strengthening certain circuitry and weakening other circuitry.

Changing thoughts and beliefs in the cortex may not be easy, but it's much easier than changing the defense response created by the amygdala. Anything we can do in the cortex to prevent the amygdala from creating the defense response and anxiety can be helpful. If you understand the connection between your cortex-based thoughts and activation of your amygdala, and recognize the amount of anxiety you can avoid by changing your thoughts, you'll be motivated to work on helping your cortex not provoke the amygdala! This work can have lasting results. If you change your thoughts, you can establish new patterns of responding in the cortex that reduce your anxiety. When José stopped expecting himself to do everything well on the first attempt, he found he had much less anxiety at work, and he became more capable of asking for help. He realized that others did not expect him to be perfect and were understanding of his occasional need for guidance. His anxiety level at work dropped a great deal, all because he changed his expectations for himself.

You can reduce not just cortex-based anxiety, but also the effects of amygdala-based anxiety, by changing your thoughts. Often the cortex worsens amygdala-based anxiety by interpreting the anxiety to mean more than it does. When Mary says, "I'm so shaky, I'm sure my presentation is going to be a disaster!" she's worsening the situation by assuming the reactions caused by her defense response predict a poor performance. Her cortex has produced thoughts interpreting the meaning of her defense response that increase her anxiety further. Changing one's thoughts is not easy, especially if you have OCD. As we noted, the particular wiring in the frontal lobes in the OCD brain makes you vulnerable to getting stuck in worry circuits, so shifting to other thoughts is not easy. But you strengthen this ability to shift each time you practice changing the channel in your cortex to new thoughts. This chapter helps you identify the amygdala-activating thoughts you want to watch for so that you can use cognitive restructuring methods to practice and strengthen your ability to change the channel.

Changing Interpretations

Because anxiety can occur due to the amygdala without any input from the cortex, changing thoughts can't always prevent anxiety. However, when thoughts or images in the cortex have activated the amygdala and led it to create the defense response, changing those thoughts can definitely reduce or prevent anxiety. Sometimes the cortex can interpret a situation in such a way that the *interpretation* causes the amygdala to be activated. (See figure 3 in chapter 3 for a reminder.) If you want to keep the cortex from activating the amygdala, change the interpretation.

When Joan hears that an unexpected meeting is scheduled for this afternoon in the office, she quickly checks her memory to recall whether she has done anything wrong and begins to wonder whether she will be reprimanded or even fired at the meeting. As a result, Joan has a great deal of anxiety as she waits for the meeting. In contrast, most people in the office would not be worried about being fired just because an unexpected meeting was scheduled, and they are not experiencing a high level of anxiety. Note that Joan's *interpretation* of the unexpected meeting leads to anxiety, not the unexpected meeting itself. If Joan changes her interpretation of the meeting by refraining from jumping to unfounded conclusions about the purpose, she won't be as anxious all day.

When you have OCD, recognizing that it's your interpretation of a situation—and not the situation itself—that often activates your amygdala is a very important bit of self-knowledge. Observing your interpretations of situations and trying to refrain from "borrowing trouble" by jumping to unfounded conclusions is very helpful in reducing amygdala activation. Why put yourself through trouble that never occurs? If you find yourself having interpretations that unnecessarily provoke the amygdala, you can try to replace those interpretations. There are a variety of ways to interpret a situation, a person's behavior, and so on. A good rule of thumb is "Don't limit yourself to a specific interpretation until you have a clear reason and sufficient evidence to rule out all other interpretations."

A common problematic interpretation is assuming that you know what another person is thinking. This is especially common in the age of texting, when we often can't correctly judge what a person is feeling from a text (even with an emoji). When Niko asked her sister if she was going to get Ho Ping a present for his birthday, her sister texted back, "Of course I am," and Niko wondered whether her sister was angry at her. Niko thought her sister could be thinking she was being nosy or that it was a stupid question. She felt very awkward and anxious about texting with her sister any more that day. So many times, even when we aren't texting, we incorrectly interpret what others are thinking about us, which can lead to a great deal of anxiety when no evidence suggests our interpretation is correct. Many times obsessing is focused on trying to mind read what is happening in another person's head. Try to resist the tendency to make interpretations about what others are thinking, which can lead to misunderstandings as well as anxiety. Remember what they say about assuming: When you ASSUME, you can make an ASS of U and ME.

Stamping Out Self-Defeating Beliefs

In this section we'll examine several common types of thoughts from the cortex that frequently activate the amygdala. Some people call these *self-defeating beliefs* because they set you up to become anxious and overwhelmed by your life experiences. They tend to increase your anxiety if you believe them. In other words, they lead to the kind of thoughts that cause the amygdala to react as if you are in danger.

Some of the self-defeating beliefs discussed below are marked with an * because these specific beliefs have been identified as promoting the development of obsessions (Fergus and Wu 2010). This is because when an intrusive thought is combined with one of these self-defeating beliefs, the intrusive thought tends to be intensified, making it more difficult to dismiss it as noise. If you have adopted these beliefs, an intrusive thought will take on more importance to you and will be more likely to become an obsession.

Consider whether you commonly find yourself thinking these thoughts or adhering to these beliefs. You have two good reasons to question these beliefs. First, these thoughts in your cortex may be activating your amygdala and increasing your anxiety. Second, they may be increasing your obsessing. If you can begin to question, challenge, and replace these beliefs with different thoughts and beliefs, you're likely to experience less anxiety and fewer obsessions.

*Perfectionism. Anxiety can be provoked by the cortex if you have perfectionistic expectations of yourself or others. Placing unrealistically high standards on yourself (or others) is guaranteed to increase your anxiety. Because no one is capable of perfection, high standards often mean you are setting expectations for yourself that can lead to frustration and disappointment. Sometimes it's clear that people learned perfectionistic expectations from others, often their parents. Parents may have good intentions when they tell their children, "Always do your best," but this exhortation can create unrealistic expectations. We simply can't be our best at every moment. If you try to have the best breakfast, make your bed the best, dress your best, brush your teeth the best, and so on, you will be exhausted before noon. This is not to say parents can't have high expectations for their children. They just need to be cautious about instilling unrealistic expectations because none of us can succeed at everything.

Parents are not always the source of perfectionistic tendencies. Some children seem to expect perfect performance of themselves from an early age, despite having parents who reassure them that they don't expect such performance. Jason said that his parents were very accepting and reasonable, but he always felt he had to do everything correctly and was extremely distressed by any mistake he made. He wasn't sure why, but he knew the perfectionism came from himself.

Some people believe that their perfectionistic expectations are reasonable, but they may not recognize that the pressure of perfection will always provoke anxiety. The self-criticism and disappointment that result from perfectionism can increase how much anxiety you experience each day. There's a big difference between believing *I need to get that assignment done* and *I need to get that assignment done with no errors*. If you look at some of your expectations for yourself, you may find perfectionism at the root of them. Fortunately, the cortex is able to set more realistic expectations, and amygdala activation will decrease as a result.

Perfectionism increases the likelihood of obsessing because it leads to people constantly judging their thoughts and behaviors against rigid standards. Perfectionism can lead to difficulty getting past a recent mistake, leading to obsessive thoughts focused on the mistake. In addition, perfectionism can lead to obsessing about upcoming performances or responsibilities, because maintaining perfection in every situation makes future events a threat. Read the statements below and check any that apply to you:

- ☐ I have high standards for myself and usually hold myself to them.

- ☐ I usually have a right way to do something and find it difficult to vary from that approach.

- ☐ People consider me an extremely conscientious and careful worker.

- ☐ When others are watching me, I'm concerned that I'll embarrass myself.

- ☐ When I'm wrong, I am extremely embarrassed and ashamed.

- ☐ I almost never perform at a level that I'm completely satisfied with.

- ☐ I have a hard time letting go of mistakes I make.

- ☐ I feel I have to be hard on myself or I won't be good enough.

Catastrophizing. When you catastrophize, you turn a small setback into a catastrophe. Although this problem is common in people with OCD, it has not been shown to lead to obsessing thinking. But it is clearly linked to provoking the amygdala to produce anxiety, so it's important for you to recognize and work on reducing it. Catastrophizing is making a mountain out of a molehill. For example, feeling that your whole day is ruined when one thing goes wrong is a cortex-based interpretation that can lead to a great deal of anxiety. Catastrophizing is influenced by the circuitry of the orbitofrontal cortex, which is an area of the brain involved in estimating the costs or downsides of events (Grupe and Nitschke 2013). Once you recognize the way catastrophizing leads to amygdala activation, you can take steps to reduce it by reminding yourself, *This is not the worst thing that could happen.* Read the statements below and check any that apply to you:

☐ Certain thoughts that go through my mind are really overwhelming.

☐ I often feel as if I can't handle one more thing going wrong.

☐ If I find a lump or a dark mark on my skin, my first thought is cancer.

☐ When I'm thinking about how some situation will turn out, I imagine the worst.

☐ When something doesn't turn out the way that I want it to, I find it difficult to cope.

☐ I overreact to problems that others wouldn't consider much of a concern.

☐ Even a small setback, like being stopped by a traffic light, can infuriate me.

☐ When something goes wrong, I'm sure there is more trouble to come.

*Misperceiving probabilities. In comparison to catastrophizing, which is judging a problem to be much worse than it is, misperceiving probabilities is about judging how *likely* it is that a problem will happen. When estimating the probability that negative events will occur, people with OCD have a tendency to overestimate their likelihood (Hezel and McNally 2016). This tendency, also called the *tendency to overestimate threat*, makes a person more pessimistic, more prone to expect bad things to happen, and more likely to activate the amygdala than those without this tendency. If Reggie thinks it's likely that the romaine lettuce is contaminated, he will be more anxious than other patrons about going out to a restaurant.

When people with OCD misperceive probabilities, they're usually focused on what *could* happen, rather than what is *likely* to happen. It doesn't matter whether the chance is 1% or 60%, they tend to think the occurrence is possible and that it poses a concern. This tendency increases the likelihood of obsessions because it makes many events seem more likely and therefore more worthy of attention, consideration, and worry (Fergus and Wu 2010). A person can become focused on events that seem extremely unlikely to most people, and the amygdala gets more activated as a result. Read the statements below and check any that apply to you:

- ☐ If I feel any physical discomfort, I usually become concerned I may be sick.

- ☐ I generally believe that, in my life, if something can go wrong, it will.

- ☐ If I lose something somewhere, I am pretty confident it is lost for good.

- ☐ I believe the world is a dangerous place.

- ☐ I frequently prepare myself for negative occurrences that don't happen.

- ☐ When I do something incorrectly at work, I worry it may lead to my being fired.

- ☐ I think it makes sense to repeatedly check anything that may cause a fire.

- ☐ Even if side effects from a medicine are rare, I expect to get them.

*Need for certainty/problems with doubt. Many people are extremely distressed by uncertainty, whether it affects their schedule, their expectations of others, or what is likely to happen in a situation. When they have a doubt about what is going to happen and don't know for sure what to expect, they experience a great deal of anxiety. If you ask most people whether they were sure they locked their houses this morning, they might say that they are 90% sure that they did, but this would not be enough certainty for those who have problems with doubt. Because life is not certain, and we often can't predict what's going to occur on a daily basis, having a need for certainty can cause the amygdala to become activated quite regularly.

Striving for certainty has been shown to lead to an increase in obsessions (Fergus and Wu 2010). A person who needs certainty and has problems tolerating doubt is likely to keep thinking about situations that are uncertain and to want to find a way to resolve the uncertainty. Because we so often find ourselves having to wait to see how situations will turn out, people who have a need for certainty can spend a lot of time obsessing about upcoming situations, checking, or seeking reassurance. Read the statements below and check any that apply to you:

☐ I like to have activities planned according to a routine schedule.

☐ If I am not sure whether I did something, I feel I should do it again to be sure.

☐ I think it's wise to consult many people before making a decision.

☐ I feel the need to check things to make sure I have done them correctly.

☐ I dislike unpredictable situations.

☐ I don't like questions that can be answered in many different ways.

☐ I don't like to try something new unless I know what to expect.

☐ I find too many options can be overwhelming.

*Guilt and responsibility. Guilt involves a feeling that you have behaved in a way that you find unacceptable or that violates a personal value you hold. Responsibility is when you feel that you are accountable for something that occurred. Both involve holding yourself to a certain standard. When a person is prone to thinking about their guilt or responsibility for something unacceptable, these thoughts are very likely to provoke activation in the amygdala. Furthermore, a person with this tendency is very vulnerable to obsessions. If one has intrusive thoughts about harm, along with a strong sense of guilt and responsibility, a harm-related thought is likely to lead to obsessive thoughts (Fergus and Wu 2010). When Tonya read the statistics about how many children develop peanut allergies, she felt she should stop working in the ice cream parlor until she was sure her infant daughter was not allergic to peanuts. She couldn't stop thinking about her guilt about working in a place that she thought could put her daughter in danger. In the same way, a person with a strong sense of guilt may be more prone to obsess about the thought of hurting someone's feelings. Read the statements below and check any that apply to you:

- ☐ It's easy for others to guilt-trip me into doing something for them.

- ☐ When someone is upset, I often suspect it is my fault.

- ☐ I review past experiences carefully to make sure I didn't do anything wrong.

- ☐ I am sometimes afraid I have accidentally harmed someone without knowing it.

- ☐ I have a hard time living with myself if I let someone down.

- ☐ When something goes wrong, I often worry that I caused it somehow.

- ☐ If I hurt someone's feelings, I can't stop thinking about it.

- ☐ I try to be very cautious with my words so that I don't accidentally offend anyone.

The cortex is a busy, noisy place, often full of ideas and expectations that have no basis in reality. The problem is not that we have all these ideas. The problem is that *we tend to take the thoughts in our cortex too seriously.* You really should not trust your cortex. Developing the ability to step back from your thoughts and not always take them at face value is very helpful. Because your amygdala often responds to thoughts from the cortex in the same way it does to actual events, you may be able to greatly reduce your anxiety and your OCD by being aware of amygdala-activating thoughts and reducing the time you spend contemplating them. Although this sounds logical, many people worry that they must take every thought or belief that they have seriously, and some even argue that the mere existence of a thought means it's true.

We rely on the cortex to process our sensations, to identify our emotions, and to allow us to perceive and think about the world. The cortex also allows us to reflect on past experiences and imagine the future. This can make it difficult to remember that the information you experience in your cortex isn't the same as reality. For example, you may think that what you remember seeing during a robbery was completely accurate, but we know from court trials that eyewitness accounts are notoriously erroneous. It's not so much that people can't always be trusted to tell the truth; it's that the cortex can't always be trusted to provide the truth. We often say that seeing is believing, but often what we see is not accurate. We've created a downloadable PowerPoint presentation, "You Can't Trust Your Cortex," available at http://www.newharbinger.com/47186, that illustrates why you should question what your cortex tells you. In this presentation you will learn that your cortex often makes you see things that aren't there, keeps you from seeing things that are clearly visible, and makes something that is nonsense into something meaningful. So check it out!

Reducing or Replacing Your Amygdala-Activating Thoughts

If you go back and review some of the self-defeating beliefs that you learned about in this chapter, you may note some beliefs that you have a tendency to agree with. If your cortex is producing thoughts that reflect these beliefs, don't let it run rampant! As Grupe and Nitschke (2013, 490) pointed out, a person with an anxiety-based disorder like OCD "probably builds up neural pathways of anxiety just as a concert pianist strengthens neural pathways of musicianship—through hours of daily practice." You can change the circuitry in your cortex by replacing self-defeating thoughts with more beneficial coping thoughts or by shifting your focus to other channels. This is the way to change the circuitry in your cortex. Many excellent self-help books are devoted to the topic of addressing dysfunctional thinking, so we won't give you detailed instructions for all of the strategies. We recommend you check out some of the books in the Resource List available at http://www.newharbinger.com/47186. Here we will describe cognitive restructuring techniques you can use to address the self-defeating beliefs that we identified earlier.

Cognitive restructuring techniques give you the power to actually change your cortex. The key is to be skeptical of your self-defeating thoughts, especially those that are amygdala-activating. Is there evidence to support these thoughts? If not, why are you putting yourself through the stress and anxiety they cause? To change these thoughts in your cortex, you can dispute them with evidence, ignore them and focus on other ideas, or replace them with new, more accurate thoughts. Pay particular attention to the amygdala-activating thoughts you catch yourself using quite often. Because several self-defeating beliefs have been shown to increase obsessions, look for connections between your obsessions and those self-defeating beliefs. As you change or replace self-defeating beliefs, you may find that your obsessing decreases.

Remember, like muscle, neural circuitry is strengthened by using it and weakened by not using it. So the more you think certain thoughts, the stronger they become, and the less attention you give certain thoughts, the weaker they become. If you replace thoughts that tend to activate the amygdala with new thoughts, you won't be activating the amygdala so frequently. You will have rewired your brain to produce less anxiety, and there will be less fuel to motivate obsessions and compulsions.

Another way to replace your amygdala-activating thoughts is with *coping thoughts*—thoughts or statements that are likely to have positive effects on a person's emotional state. One way of evaluating the usefulness of thoughts is to look at the effects they have on you. In this light, you can clearly see the value of coping thoughts, which are more likely to result in calm responding and an increased ability to cope with difficult situations. They are less likely to activate the amygdala. Here are some examples.

Self-Defeating Thought (Amygdala-Activating)	Coping Thought (Calming)
I must be competent and excel in everything I do.	*No one is perfect; I will make mistakes and it is okay.*
This situation is going to turn into a disaster.	*I don't know what is going to happen, so I won't panic until I have a reason.*
I won't be able to handle this.	*I've been through worse than this!*
There is a good chance that this will turn out badly.	*I'm just stressing myself out with these thoughts; I don't know what is going to happen.*
I'm sure that the worst will happen.	*My OCD always exaggerates the danger and I need to focus on living my life today.*
I need to know what's going to happen; I can't tolerate this uncertainty.	*I need this opportunity to live with uncertainty because it's part of life and I want to practice living with it.*
I don't know for sure whether this will turn out all right.	*I can handle not knowing; I'll deal with problems if they come.*
I may have harmed/offended someone.	*This is a thought and I don't need to focus on it without any evidence for it.*
I need to check to make sure. What if I did something wrong?	*This is a good chance to practice letting go and living my life.*

Of course, you'll have to be vigilant about recognizing amygdala-activating thoughts and substituting coping thoughts, but it's worth the effort. Some people post their coping thoughts around their living spaces as reminders. By deliberately thinking coping thoughts at every opportunity, you can rewire your cortex to produce coping thoughts on its own. The more you use them, the more they pop into your head. At first, they may feel awkward, but they become more familiar and comfortable as you use them.

While working on changing your thoughts, you may feel that you can't get rid of negative thoughts you try to eliminate. This is a common problem that springs from how the brain works. As we have noted, you can't weaken circuitry by using it. Studies have shown that trying to erase, silence, or turn off a thought simply is not an effective approach (Wegner et al. 1987). Right now, if I tell you, "Don't think about pink elephants," you are going to have pink elephants come into your mind. By asking you not to think about them, I activated the circuitry storing pink elephants in your cortex. This is important to understand because in order to try to stop thinking about something, like your obsessions, you need to understand that telling yourself not to think about something just will not work. Trying to erase a thought by constantly reminding yourself not to think about it (and therefore thinking about it) activates the circuitry storing the thought and just makes the thought stronger.

You might be successful in interrupting a thought by saying "Stop!" to yourself when you catch yourself thinking it. This technique is known as *thought stopping*. However, the next step is crucial. If you then *replace* the thought with another thought, it's more likely that you will keep the first thought out of your mind. For example, as Trevor is working in his garden, he looks back and sees that he has not planted perfectly straight rows. The thought instantly leads to anxiety as his amygdala reacts to the thought, *I have made a mistake!* Knowing he is trying not to be controlled by his perfectionism, Trevor says "Stop!" but does not argue with the thought. He turns around and looks at all the tools, empty containers, and fertilizer, and says to himself, *I need to get this mess cleaned up and get ready for dinner. What should I make?* By replacing the amygdala-activating thought with something else that engages your mind, you are less likely to return to that thought.

Therefore, "Don't erase—replace!" is the best approach with amygdala-activating thoughts. If you notice that you're thinking something you want to shift away from, like Trevor did, focus on replacing that thought with a new one. By stating a different thought to yourself to replace the first one, you will strengthen a more adaptive way of thinking and activate circuitry that will help you resist anxiety. Remember, we can't focus on two thoughts at the same time. It takes some practice to resist shifting back to the first thought, but eventually changing your focus to new thoughts becomes habitual. The use of coping thoughts is an ideal way to apply the process of replacing an amygdala-activating, self-defeating thought with one that is less likely to provoke anxiety and more helpful in allowing you to live the life you want.

Calming Your Cortex Through Acceptance

While cognitive restructuring techniques can modify your thinking and minimize amygdala activation, we need to reemphasize that, by itself, a thought does not pose a danger. What it can do is provoke a response from the amygdala, which elicits the defense response and anxiety. We want to help you understand and interrupt the patterns of responding in your cortex so that your brain is less likely to produce the anxiety that fuels your obsessing and compulsions. We are helping you rewire your cortex so that it does not repetitively provoke anxiety and other symptoms. We are not trying to eliminate "dangerous thoughts"; we are trying to prevent unnecessary amygdala activation.

But the truth is, you don't have to control amygdala responding. You can allow the amygdala to respond and use mindfulness to observe it without getting caught up in the response. This is another way to manage your cortex—by taking a mindful approach to anxiety, focusing on observing the amygdala's response with curiosity and acceptance. When you do this, the cortex gives up the goal of controlling the situation and simply allows anxiety to happen. You experience and observe your emotional reaction with curiosity and acceptance. This acceptance of your experience is another antidote to anxiety. People who use mindful approaches have an amygdala that is generally less activated.

Acceptance of the bodily reactions (often called symptoms) that the amygdala creates can also make a big difference in your life. These reactions do not pose any danger or signal any upcoming difficulty. Accepting the pounding heart, trembling hands, muscle tension, or feeling of dread with patience, understanding, or even humor can change the impact of the amygdala on your life. To accurately describe the way the amygdala operates, say to yourself, *My silly amygdala thinks I am in some kind of danger when my desk is in disarray and gets my heart pounding as if I should run away from my office.* If you can stop acting as though your anxiety means your desk is a problem, you are less likely to obsess about it. After you recognize the influence of the amygdala, acceptance and coping strategies can help you avoid misinterpretations and obsessive thoughts that just worsen your OCD.

Much of the distress of OCD comes from the constant struggle to fight thoughts and manage anxiety. If you take the approach of accepting thoughts that come into your cortex and accepting the experience of anxiety that can follow, you are disengaging from the struggle. In this approach, you allow thoughts without attempting to control them, remembering that they are simply thoughts and observing them rather than being caught up in them. Practice letting go of trying to control your thoughts; just let them flow or bounce around, whatever they do, and simply be observant and curious about them. If anxiety arises, let yourself experience it and explore what it is like. When you accept and observe your anxiety, being aware of different aspects of how it is experienced in your body, it loses the ability to control you. Knowing that anxiety will pass eventually and accepting it will actually cause it to pass more quickly. You won't perpetuate it with a fearful reaction to it. Much of the discomfort of anxiety arises from struggling with it and trying to force it to go away. Strange as it seems, by giving up attempts to control your thoughts and your anxiety, you can actually be more in control of your brain.

Studies show amazing changes in the brains of people who practice mindfulness and other forms of meditation. They are able to more quickly reduce their anxiety in the present moment (Zeidan et al. 2013) and experience lasting changes in the cortex that make them more resistant to anxiety. Those who are experienced in mindfulness don't necessarily show changes in the amygdala; the neural images show that they've disengaged the cortex from getting caught up in the amygdala's responding in the way it usually does (Froeliger et al. 2012). With mindfulness, you train the cortex to respond to anxiety in a completely new way. Neuroimaging studies show that the few parts of the cortex that have a direct connection to the amygdala—the ventral medial prefrontal cortex and the anterior cingulate cortex—are the very parts of the cortex activated by mindfulness meditation (Zeidan et al. 2013). These findings indicate that mindfulness approaches can help you rewire parts of the cortex that are intimately connected with calming the amygdala.

Mindfulness training can change the way your cortex responds to anxiety. Making mindfulness part of your daily life can change your relationship with your cortex and amygdala and help you manage your OCD better. We recommend that you explore the usefulness of mindfulness in coping with OCD. Many excellent books and video resources can provide training in mindfulness, and the ones that focus specifically on anxiety or OCD will be most useful. (See the Resource List at http://www.newharbinger.com/47186 for recommendations.)

In this chapter, you learned several approaches to helping your cortex respond to anxiety in new ways: using cognitive restructuring techniques, changing your interpretations of situations, identifying how your cortex creates anxiety, reducing or replacing (not erasing!) thoughts that activate your amygdala, and finally, using acceptance strategies to calm your cortex. As you use these approaches to rewire your cortex, you will be increasingly able to resist the control of OCD, move past the limits that OCD imposes on you, and live the life YOU want to live. The last step is to put everything you have learned in this book together, so please read on for some final thoughts that can help you do that.

Conclusion

Our goal in this book has been to provide you with knowledge of the brain processes that are involved in OCD, and we hope that the information we've shared with you will help you live the life you want to live, despite OCD. Obsessions and compulsive behaviors do not need to dominate your life if you understand how to rewire your obsessive brain so that you have effective ways to manage your obsessions and your anxiety. We are certain that your new understanding of what occurs in your brain to give OCD control over your life is nothing you expected when you first picked up this book. You have learned methods that can help put you back in control of your life—methods that can be effective no matter what the content of your obsessions or the nature of your compulsions.

Having spent time trying to get rid of thoughts and obsessions that torment you and trying to find ways to survive your anxiety, you are now in a new position. Understanding how and why the defense response is created in the amygdala and recognizing the way in which the cortex contributes to activating the amygdala has helped you understand the true source of the anxiety that you experience. Hopefully, you recognize the usefulness of the Serenity Prayer: you have to focus on what you can control, and try to control that, but accept that some things can't be controlled. You have learned that the defense response and your feelings of anxiety are not completely within your conscious control, and also that, while distressing, your anxiety is not a danger. You can't change that your brain is designed to produce anxiety, but you can understand it better and learn to manage it differently.

Similarly, you have learned that you can't always trust your cortex and that thoughts that arise in your cortex, including your worries and obsessions, are not dangerous, even though they sometimes activate the amygdala. You have learned strategies for managing your cortex, but have also learned that no one can expect to have complete control of the thoughts that come up in the cortex. By explaining the neuroplasticity of the brain and the methods required for teaching your amygdala and cortex to respond differently, we have shown you strategies you can use to rewire your brain to resist obsessing and acting on compulsions. These strategies will allow you to overcome the limits of OCD that keep you from living the life you want.

Setbacks will certainly occur in the process of coping with OCD. A setback is only a loss if you give up. This is your life, and OCD should not be in charge, so don't give up and give control back to OCD. Of course, ships are safer when they stay in the harbor, but they aren't meant to stay there. Push through fear in order to be the master of your own life. The key to managing OCD is to focus on following your own goals and not let anxiety and fear stop you. All the knowledge you have gained from this book can help you manage your OCD more effectively and gradually rewire your brain to find more satisfaction in your life. We hope this journey brings you relief, encouragement, satisfaction, and joy. You deserve it!

We don't expect you to take on your OCD alone. We encourage you to find a therapist who understands how to help you with the cognitive behavioral approaches that have been shown to be so helpful in overcoming OCD, especially ERP and cognitive restructuring. Because you've read this book so thoroughly, you may be hoping to avoid working with a therapist. Perhaps you are reluctant to do so because your thoughts and behaviors are so personal. We want you to know that therapists experienced in treating OCD will not be shocked by your thoughts or behaviors and can help you with them if you open yourself up to their assistance. Everything you discuss will be confidential. If you have had an unsatisfactory experience with a therapist, please consider that you just have not found the right person yet, and keep looking. In addition, you may decide to consult a psychiatrist or other physician about medications that could assist in your treatment. If so, please read the bonus chapter "Are Medications Needed in the Obsessive Brain?" available at http://www.newharbinger.com/47186.

We also encourage you to seek training in mindfulness because it will enable you to focus on your experiences in new ways. Mindfulness can give you an ability to take your focus off your obsessions, providing a level of freedom from OCD that is hard to imagine. Resources are available in many communities and in apps to provide mindfulness training.

Your new knowledge of the role of the amygdala and your awareness of the influences of the cortex are valuable tools in the process of taking back your life from the influences of OCD. You have a new understanding of the nature of anxiety along with many different methods to manage the amygdala and your anxiety. We have also introduced you to various techniques that you can use to rewire your cortex so that obsessions, worries, and self-defeating beliefs don't dominate your thinking and constrict your life. You can make lasting changes in your brain instead of living each day to reduce anxiety temporarily through obsessions and compulsions that keep you in the endless cycle of OCD. We hope you will use this information to make new connections in your brain that allow you the freedom to reclaim your life and live each day more in accordance with your own goals and values, not restricted by the unreasonable demands of OCD.

Acknowledgments

From Catherine: As I consider those I appreciated during the process of writing this book, I first think of many clients who have shared with me their struggles with OCD, anxiety, and worries; they have had the courage to push through the feelings and thoughts that could have blocked them from living the lives they wanted. I feel privileged to have learned so much from all of you, and I admire your ability to continue working each day on living the Serenity Prayer in your lives. Next, I thank my wife, Vickie, for her patience and encouragement every day that she had to watch me sitting at the computer, and I appreciate the many hours she sat beside me (along with our faithful dogs) as I worked on the manuscript. I also want to thank my daughters, Arrianna and Melinda, for their encouragement and understanding when I am so busy with my work. Special thanks to Arri for all the assistance she provided with the illustrations for the book and all she does with social media to help me get messages of hope and encouragement out to those trying to manage their anxiety. Without her, I would not be found on Twitter and Instagram @drcmpittman. Finally, thanks to Bill Youngs for sticking with me through this project despite health issues, pandemic, quarantine, and a totally unacceptable lack of lunch meetings.

From Bill: Any intellectual accomplishment is ultimately the result of multiple influences in one's life. In this regard, I want to acknowledge the influences of my clinical and academic colleagues at Memorial Hospital of South Bend, Indiana; Psychology Associates of Mishawaka, Indiana; and Saint Mary's College in Notre Dame, Indiana; as well as the influence of the many patients and clients I have seen over the years. You have all shaped my clinical understanding and compassion for others in so many ways.

Catherine Pittman and I have been colleagues and friends for over thirty years and have consulted with one another on both clinical and academic issues as well as the topics that good friends discuss. A very important part of our friendship has been our weekly lunch meetings, where we discuss both minor and major issues in our lives. Unfortunately, these meetings have been put on hold because of the global health crisis we face, but I look forward to being able to meet for lunch again.

Finally, I want to give my greatest thanks to my wife, Diane, for her encouragement and patience with me and my work over the years. She is the love of my life and has always been there for me. We are a team and have been through so much together.

References

American Psychiatric Association. 2013. *Diagnostic and Statistical Manual of Mental Disorders 5*. Washington, DC: American Psychiatric Association.

Anderson, E., and G. Shivakumar. 2013. "Effects of Exercise and Physical Activity on Anxiety." *Frontiers in Psychiatry* 4: Article 27.

Andrzejewski, J. A., T. Greenberg, and J. M. Carlson. 2019. "Neural Correlates of Aversive Anticipation: An Activation Likelihood Estimate Meta-Analysis Across Multiple Sensory Modalities." *Cognitive, Affective, and Behavioral Neuroscience* 19: 1379–1390.

Asan, E., M. Steinke, and K. Lesch. 2013. "Serotonergic Innervation of the Amygdala: Targets, Receptors, and Implications for Stress and Anxiety." *Histochemistry & Cell Biology* 139: 785–813.

Baj, J., E. Sitarz, A. Forma, et al. 2020. "Alterations in the Nervous System and Gut Microbiota After B-Hemolytic Streptococcus Group A Infection—Characteristics and Diagnostic Criteria of PANDAS Recognition." *International Journal of Molecular Science* 21: 1476–1500.

Beck, A. T. 1976. *Cognitive Therapy and the Emotional Disorders*. New York: Penguin Group.

Bernacer, J., I. Martinez-Valbuena, M. Martinez, et al. 2019. "An Amygdala-Cingulate Network Underpins Changes in Effort-Based Decision Making After a Fitness Program." *Neuroimage* 203: ArtID116181.

Bernstein, G. A., A. M. Victor, A. J. Pipal, and K. A. Williams. 2010. "Comparison of Clinical Characteristics of Pediatric Autoimmune Neuropsychiatric Disorders Associated with Streptococcal Infections and Childhood Obsessive-Compulsive Disorder." *Journal of Child and Adolescent Psychopharmacology* 20: 333–340.

Bonnet, M. H. 1985. "Effect of Sleep Disruption on Sleep, Performance, and Mood." *Sleep* 8: 11–19.

Bourne, E. J., A. Brownstein, and L. Garano. 2004. *Natural Relief for Anxiety: Complementary Strategies for Easing Fear, Panic, and Worry*. Oakland, CA: New Harbinger.

Cannon, W. B. 1929. *Bodily Changes in Pain, Hunger, Fear, and Rage*. New York: Appleton.

Chen, Y., C. Chen, R. M. Martinez, J. E. Etnier, and Y. Cheng. 2019. "Habitual Physical Activity Mediates the Acute Exercise-Induced Modulation of

Anxiety-Related Amygdala Functional Connectivity." *Scientific Reports* 9: 19787.

Clark, D. A. 2020. *Cognitive-Behavioral Therapy for OCD and Its Subtypes*, 2nd ed. New York: The Guilford Press.

Cotman, C. W., and N. C. Berchtold. 2002. "Exercise: A Behavioral Intervention to Enhance Brain Health and Plasticity." *Trends in Neurosciences* 25: 295–301.

Craske, M. G., and D. H. Barlow. 2007. *Mastery of Your Anxiety and Panic: Therapist Guide*. 4th ed. New York: Oxford University Press.

Davidson, J. 2014. *Daring to Challenge OCD: Overcome Your Fear of Treatment and Take Control of Your Life Using Exposure and Response Prevention.* Oakland, CA: New Harbinger.

Davidson, R. J., and S. Begley. 2012. *The Emotional Life of Your Brain: How Its Unique Patterns Affect the Way You Think, Feel and Live—And How You Can Change Them.* New York: Hudson Street Press.

DeBoer, L., M. Powers, A. Utschig, M. Otto, and J. Smits. 2012. "Exploring Exercise as an Avenue for the Treatment of Anxiety Disorders." *Expert Review of Neurotherapeutics* 12: 1011–1022.

de Salles Andrade, J. B., F. M. Ferreira, C. Suo, et al. 2019. "An MRI Study of the Metabolic and Structural Abnormalities in Obsessive-Compulsive Disorder." *Frontiers of Human Neuroscience* 13: Article 186.

Desbordes, L. T., T. W. W. Negi, B. A. Pace, C. L. Wallace, C. L. Raison, and E. L. Schwartz. 2012. "Effects of Mindful-Attention and Compassion Meditation Training on Amygdala Response to Emotional Stimuli in an Ordinary, Non-Meditative State." *Frontiers in Human Neuroscience* 6: 1–15.

Dias, B., S. Banerjee, J. Goodman, and K. Ressler. 2013. "Towards New Approaches to Disorders of Fear and Anxiety." *Current Opinion in Neurobiology* 23: 346–352.

Doidge, N. 2007. *The Brain That Changes Itself: Stories of Personal Triumph from the Frontiers of Brain Science.* New York: Penguin.

Doll, A., B. K. Holzel, S. M. Bratec, et al. 2016. "Mindful Attention to Breath Regulates Emotions via Increased Amygdala-Prefrontal Cortex Connectivity." *NeuroImage* 134: 305–313.

Drew, M. R., and R. Hen. 2007. "Adult Hippocampal Neurogenesis as Target for the Treatment of Depression." *CNS & Neurological Disorders-Drug Targets* 6: 205–218.

Ellis, A., and K. Doyle. 2016. *How to Control Your Anxiety Before It Controls You.* New York: Citadel.

Engels, A. S., W. Heller, A. Mohanty, J. D. Herrington, M. T. Banich, A G. Webb, and G. A. Miller. 2007. "Specificity of Regional Brain Activity in

Anxiety Types During Emotional Processing." *Psychophysiology* 44: 352–363.

Ensari, I., T. A. Greenlee, R. W. Motl, and S. J. Petruzzello. 2015. "Meta-Analysis of Acute Exercise Effects on State Anxiety: An Update of Randomized Controlled Trials over the Past 25 Years." *Depression and Anxiety* 32: 624–634.

Fergus, T., and K. D. Wu. 2010. "Do Symptoms of Generalized Anxiety and Obsessive-Compulsive Disorder Share Cognitive Processes?" *Cognitive Therapy & Research* 34: 168–176.

Foa, E. B., J. D. Huppert, and S. P. Cahill. 2006. "Emotional Processing Theory: An Update." In *Pathological Anxiety: Emotional Processing in Etiology and Treatment*, edited by B. O. Rotham. New York: Guilford.

Freeston, M. H., M. J. Dugas, and R. Ladouceur. 1996. "Thoughts, Images, Worry, and Anxiety." *Cognitive Therapy and Research* 20: 265–273.

Froeliger, B. E., E. L. Garland, L. A. Modlin, and F. J. McClernon. 2012. "Neurocognitive Correlates of the Effects of Yoga Meditation Practice on Emotion and Cognition: A Pilot Study." *Frontiers in Integrative Neuroscience* 6: 1–11.

Fullana, M. A., X. Zhu, P. Alonso, et al. 2017. "Basolateral Amygdala-Ventromedial Prefrontal Cortex Connectivity Predicts Cognitive Behavioral Therapy Outcome in Adults with Obsessive-Compulsive Disorder." *Journal of Psychiatry Neuroscience* 46: 378–385.

Gibbs, N. A. 1996. "Nonclinical Populations in Research on Obsessive-Compulsive Disorder: A Critical Review." *Clinical Psychology Review* 16: 729–773.

Gloster, A. T., H. U. Wittchen, F. Einsle, et al. 2011. "Psychological Treatment for Panic Disorder with Agoraphobia: A Randomized Controlled Trial to Examine the Role of Therapist-Guided Exposure in Situ in CBT." *Journal of Counseling and Clinical Psychology* 79(3): 406–420.

Goldin, P. R., and J. J. Gross. 2010. "Effects of Mindfulness-Based Stress Reduction (MBSR) on Emotion Regulation in Social Anxiety Disorder." *Emotion* 10: 83–91.

Goleman, D. 1995. *Emotional Intelligence*. New York: Bantam.

Greenwood, B. N., P. V. Strong, A. B. Loughridge, H. E. Day, P. J. Clark, A. Mika, et al. 2012. "5-HT2C Receptors in the Basolateral Amygdala and Dorsal Striatum Are a Novel Target for the Anxiolytic and Antidepressant Effects of Exercise." *PLoS ONE* 7: e46118.

Gregory, A., and T. C. Eley. 2007. "Genetic Influences on Anxiety in Children: What We've Learned and Where We're Heading." *Clinical Child and Family Psychology* 10: 199–212.

Grupe, D. W., and J. B. Nitschke. 2013. "Uncertainty and Anticipation in Anxiety: An Integrated Neurobiological and Psychological Perspective." *Nature Reviews Neuroscience* 14: 488–501.

Hayes, S. C., V. M. Follette, and M. Linehan. 2004. *Mindfulness and Acceptance: Expanding the Cognitive-Behavioral Tradition*. New York: Guilford Press.

Hebb, D. O. 1949. *The Organization of Behavior*. New York: Wiley.

Heisler, L., K. Zhou, P. Bajwa, J. Hsu, and L. H. Tecott. 2007. "Serotonin 5-HT2c Receptors Regulate Anxiety-Like Behavior." *Genes, Brain, and Behavior* 6: 491–496.

Hertenstein, E., N. Rose, U. Voderholzer, et al. 2012. "Mindfulness-Based Cognitive Therapy in Obsessive-Compulsive Disorder: A Qualitative Study on Patients' Experiences." *BMC Psychiatry* 12: 185–195.

Hezel, D. M., and R. J. McNally. 2016. "A Theoretical Review of Cognitive Biases and Deficits in Obsessive-Compulsive Disorder." *Biological Psychology* 121: 221–232.

Hirsch, C. R., S. Hayes, A. Mathews, G. Perman, and T. Borkovek. 2012. "The Extent and Nature of Imagery During Worry and Positive Thinking in Generalized Anxiety Disorder." *Journal of Abnormal Psychology* 121: 238–243.

Hoexter, M. Q., and M. C. Batistuzzo. 2018. "Disentangling the Role of Amygdala Activation in Obsessive-Compulsive Disorder." *Biological Psychiatry* 6: 499–500.

Holzel, B. K., S. W. Lazar, T. Gard, Z. Schuman-Olivier, D. R. Vago, and U. Ott. 2011. "How Does Mindfulness Meditation Work? Proposing Mechanisms of Action from a Conceptual and Neural Perspective." *Perspectives on Psychological Science* 6: 537–559.

Jacobson, E. 1938. *Progressive Relaxation*. Chicago: University of Chicago Press.

Jasuja, V., G. Purohit, S. Mendpara, and B. M. Palan. 2014. "Evaluation of Psychological Symptoms in Premenstrual Syndrome Using PMR Technique." *Journal of Clinical and Diagnostic Research* 8: BC01–BC03.

Jerath, R., V. A. Barnes, and M. W. Crawford. 2014. "Mind-Body Response and Neurophysiological Changes During Stress and Meditation: Central Role of Homeostasis." *Journal of Biological Regulators and Homeostatic Agents* 28: 545–554.

Jerath, R., V. A. Barnes, D. Dillard-Wright, S. Jerath, and B. Hamilton. 2012. "Dynamic Change of Awareness During Meditation Techniques: Neural and Physiological Correlates." *Frontiers in Human Science* 6: 1–4.

Jerath, R., V. M. W. Crawford, V. A. Barnes, and K. Harden. 2015. "Self-Regulation of Breathing as a Primary Treatment for Anxiety." *Applied Psychophysiology & Biofeedback* 40: 107–115.

Jerath, R., J. W. Edry, V. A. Barnes, and V. Jerath. 2006. "Physiology of Long Pranayamic Breathing: Neural Respiratory Elements May Provide a Mechanism That Explains How Slow Deep Breathing Shifts the Autonomic Nervous System." *Medical Hypothesis* 67: 566–571.

Johnsgard, K. W. 2004. *Conquering Depression and Anxiety through Exercise.* Amherst, NY: Prometheus Books.

Kalyani, B. G., G. Venkatasubramanian, R. Arasappa, N. P. Rao, S. V. Kalmady, R. V. Behere, H. Rao, M. K. Vasudev, and B. N. Gangadhar. 2011. *International Journal of Yoga* 4: 3–6.

Kim, T., J. Kim, J. Park, et al. 2015. "Antidepressant Effects of Exercise Are Produced via Suppression of Hypocretin/Orexin and Melanin-Containing Hormone in the Basolateral Amygdala." *Neurobiology of Disease* 79: 59–69.

Kishi, A., S. Haraki, R. Toyota, et al. 2020. "Sleep Stage Dynamics in Young Patients with Sleep Bruxism." *Journal of Sleep Research and Sleep Disorders Research* 43: 1–12.

Koran, L. M., and H. B. Simpson. 2013. *Guideline Watch (March 2013): Practice Guideline for the Treatment of Patients with Obsessive-Compulsive Disorder.* Washington, DC: American Psychiatric Association.

Lang, T., and S. Helbig-Lang. 2012. "Exposure in Vivo with and Without Presence of a Therapist: Does It Matter?" In *Exposure Therapy*, edited by P. Neudeck and H. U. Wittchen. New York: Springer.

Lattari, E., H. Budde, F. Paes, et al. 2018. "Effects of Aerobic Exercise on Anxiety Symptoms and Cortical Activity in Patients with Panic Disorder: A Pilot Study." *Clinical Practice and Epidemiology in Mental Health* 14: 11–25.

Leaver, A. M., J. Van Lare, B. Zielinski, A. R. Halpern, and J. P. Rauschecker. 2009. "Brain Activation During Anticipation of Sound Sequences." *The Journal of Neuroscience* 29: 2477–2485.

LeDoux, J. E. 1996. *The Emotional Brain: The Mysterious Underpinnings of Emotional Life.* New York: Simon & Schuster.

LeDoux, J. E. 2002. *Synaptic Self: How Our Brains Become Who We Are.* New York: Viking.

LeDoux, J. E. 2015. *Anxious: Using the Brain to Understand and Treat Fear and Anxiety.* New York: Penguin.

LeDoux, J. E., and D. Schiller. 2009. "The Human Amygdala: Insights from Other Animals." In *The Human Amygdala*, edited by P. J. Whalen and E. A. Phelps. New York: Guilford Press.

Leung, M. K., K. W. Way, C. H. Chetwyn, et al. 2018. "Meditation-Induced Neuroplastic Changes in Amygdala Activity During Negative Affective Processing." *Social Neuroscience* 13: 277–288.

Linden, D. E. 2006. "How Psychotherapy Changes the Brain—The Contribution of Functional Neuroimaging." *Molecular Psychiatry* 11: 528–538.

Lucibello, K. M., J. Parker, and J. J. Heisz. 2019. "Examining a Training Effect on the State Anxiety Response to an Acute Bout of Exercise in Low and High Anxious Individuals." *Journal of Affective Disorders* 247: 29–35.

McLean, C. P., L. J. Zandberg, P. E. VanMeter, et al. 2015. "Exposure and Response Prevention Helps Adults with Obsessive-Compulsive Disorder Who Do Not Respond to Pharmacological Augmentation Strategies." *Journal of Clinical Psychiatry* 76: 1653–1657.

Milham, M. P., A. C. Nugent, W. C. Drevets, D. P. Dickstein, E. Leibenluft, M. Ernst, D. Charney, and D. S. Pine. 2005. "Selective Reduction in Amygdala Volume in Pediatric Anxiety Disorders: A Voxel-Based Morphometry Investigation." *Biological Psychiatry* 57: 961–966.

Mochcovitch, M. D, A. C. Deslandes, R. C. Freire, R. F. Garcia, and A. E. Nardi. 2016. "The Effects of Regular Physical Activity on Anxiety Symptoms in Healthy Older Adults: A Systematic Review." *Revista Brasileira de Psiquiatria* 38: 255–261.

Nazeer, A., F. Latif, A. Mondal, M. W. Azeem, and D. E. Graydanus. 2020. "Obsessive-Compulsive Disorder in Children and Adolescents: Epidemiology, Diagnosis, and Management." *Translational Pediatrics* 9: S76–S93.

Neeru, D., C. Khakha, S. Satapathy, and A. B. Dey. 2015. "Impact of Jacobson Progressive Muscle Relaxation (JPMR) and Deep Breathing Exercises on Anxiety, Psychological Distress, and Quality of Sleep of Hospitalized Older Adults." *Journal of Psychosocial Research* 10: 211–223.

Nestadt, G., M. Grados, and J. F. Samuels. 2010. "Genetics of OCD." *Psychiatric Clinics of North America* 33: 141–158.

Nolen-Hoeksema, S. 2000. "The Role of Rumination in Depressive Disorders and Mixed Anxiety/Depressive Symptoms." *Journal of Abnormal Psychology* 109: 504–511.

Ohman, A., and S. Mineka. 2001. "Fears, Phobias, and Preparedness: Toward an Evolved Module of Fear and Fear Learning." *Psychological Review* 108: 483–522.

Olsson, A., K. I. Nearing, and E. A. Phelps. 2007. "Learning Fears by Observing Others: The Neural Systems of Social Fear Transmission." *Social Cognitive and Affective Neuroscience* 2: 3–11.

Orisillo, S. M., L. Roemer, J. B. Lerner, and M. T. Tull. 2004. "Acceptance, Mindfulness, and Cognitive-Behavioral Therapy: Comparisons, Contrasts, and Application to Anxiety." In *Mindfulness and Acceptance: Expanding the Cognitive-Behavioral Tradition*, edited by S. C. Hayes, V. M. Follette, and M. Linehan New York: Guilford Press.

Paul, S., J. C. Beucke, C. Kaufmann, et al. 2018. "Amygdala-Prefrontal Connectivity During Appraisal of Symptom-Related Stimuli in Obsessive-Compulsive Disorder." *Psychological Medicine* 49: 278–286.

Paulesu, E., E. Sambugaro, T. Torti, et al. 2010. "Neural Correlates of Worry in Generalized Anxiety Disorder and in Normal Controls: A Functional MRI Study." *Psychological Medicine* 40: 117–124.

Quirk, G. J., J. C. Repa, and J. E. LeDoux. 1995. "Fear Conditioning Enhances Short-Latency Auditory Responses of Lateral Amygdala Neurons: Parallel Recordings in the Freely Behaving Rat." *Neuron* 15: 1029–1039.

Rachman, S., and P. de Silva. 1978. "Abnormal and Normal Obsessions." *Behaviour Research and Therapy* 16: 233–248.

Radomsky, A. S., M. J. Dugas, G. M. Alcolado, and S. L. Lavoie. 2014. "When More Is Less: Doubt, Repetition, Memory, Metamemory, and Compulsive Checking in OCD." *Behaviour Research and Therapy* 59: 30–39.

Rebar, A., R. Stanton, D. Geard, C. Short, M. J. Duncan, and C. Vandelanotte. 2015. "A Meta-Meta-Analysis of the Effect of Physical Activity on Depression and Anxiety in a Non-Clinical Adult Population." *Health Psychology Review* 9: 366–378.

Roy, M. J., M. E. Costanzo, J. R. Blair, and A. A. Rizzo. 2014. "Compelling Evidence That Exposure Therapy for PTSD Normalizes Brain Function." In *Annual Review of Cybertherapy and Telemedicine*, edited by B. K. Wiederhold and G. Riva. Amsterdam, Netherlands: IOS Press.

Rupp, C., C. Jurgens, P. Doebler, F. Andor, and U. Buhlmann. 2019. "A Randomized Waitlist-Controlled Trial Comparing Detached Mindfulness and Cognitive Restructuring in Obsessive-Compulsive Disorder." *PLoS ONE* 14: e0213895.

Sander, D., J. Grafman, and T. Zalla. 2003. "The Human Amygdala: An Evolved System for Relevance Detection." *Reviews in the Neurosciences* 14: 303–316.

Sapolsky, R. M. 1998. *Why Zebras Don't Get Ulcers: An Updated Guide to Stress, Stress-Related Diseases, and Coping.* New York: W. H. Freeman and Company.

Schmitt, A., N. Upadhyay, J. A. Martin, S. R. Vega, H. K. Struder, and H. Boecker. 2020. "Affective Modulation After High-Intensity Exercise Is Associated with Prolonged Amygdala-Insular Functional Connectivity Increase." *Neural Plasticity*: Article ID 7905387.

Schmolesky, M. T., D. L. Webb, and R. A. Hansen. 2013. "The Effects of Aerobic Exercise Intensity and Duration on Levels of Brain-Derived Neurotropic Factor in Healthy Men." *Journal of Sports Science and Medicine* 12: 502–511.

Schwartz, J. M., and S. Begley. 2003. *The Mind and the Brain: Neuroplasticity and the Power of Mental Force.* New York: HarperCollins.

Shafir, T. 2015. "Movement-Based Strategies for Emotion Regulation." In *Handbook on Emotion Regulation*, edited by M. L. Bryant. Hauppauge, NY: Nova Science Publishers.

Silton, R. L., W. Heller, A. Engels, et al. 2011. "Depression and Anxious Apprehension Distinguish Frontocingulate Cortical Activity During Top-Down Attentional Control." *Journal of Abnormal Psychology* 120: 272–285.

Spoormaker, V. I., and J. van den Bout. 2005. "Depression and Anxiety Complaints: Relations with Sleep Disturbances." *European Psychiatry* 20: 243–245.

Swain, R. A., K. L. Berggren, A. L. Kerr, et al. 2012. "On Aerobic Exercise and Behavioral and Neural Plasticity." *Brain Sciences* 2: 709–744.

Taren, A. A., J. D. Creswell, and P. J. Gianaros. 2013. "Dispositional Mindfulness Co-Varies with Smaller Amygdala and Caudate Volumes in Community Adults." *PLoS ONE* 8(5): e64574.

Taylor, V. A., J. Grant, V. Daneault, et al. 2011. "Impact of Mindfulness on the Neural Responses to Emotional Pictures in Experienced and Beginner Meditators." *Neuroimage* 57: 1524–1533.

Thorsen, A. L., P. Hagland, J. Radua, et al. 2018. "Emotional Processing in Obsessive-Compulsive Disorder: A Systematic Review and Meta-Analysis of 25 Functional Neuroimaging Studies." *Biological Psychiatry Cognitive Neuroscience Neuroimaging* 3: 563–571.

van der Helm, E., J. Yao, S. Dutt, V. Rao, J. M. Salentin, and M. P. Walker. 2011. "REM Sleep Depotentiates Amygdala Activity to Previous Emotional Experiences." *Current Biology* 21: 2029–2032.

Via, E., N. Cardoner, J. Pujol, et al. 2014. "Amygdala Activation and Symptom Dimensions in Obsessive-Compulsive Disorder." *British Journal of Psychiatry* 204: 61–68.

Virkuil, B., J. F. Brosschot, T. D. Borkovec, and J. F. Thayer. 2009. "Acute Autonomic Effects of Experimental Worry and Cognitive Problem Solving: Why Worry About Worry?" *International Journal of Clinical and Health Psychology* 9: 439–453.

Vrana, S. R., B. N. Cuthbert, and P. J. Lang. 1986. "Fear Imagery and Text Processing." *Psychophysiology* 23: 247–253.

Walsh, R., and L. Shapiro. 2006. "The Meeting of Meditative Disciplines and Western Psychology: A Mutually Enriching Dialogue." *American Psychologist* 61: 227–239.

Wassing, R., O. Lakbila-Kamal, J. R. Ramautar, et al. 2019. "Restless REM Sleep Impedes Overnight Amygdala Adaptation." *Current Biology* 29: 2351–2358.

Way, B. M., J. D. Cresswell, N. I. Eisenberger, and M. D. Lieberman. 2010. "Dispositional Mindfulness and Depressive Symptomatology: Correlations with Limbic and Self-Referential Neural Activity During Rest." *Emotion* 10: 12–24.

Wegner, D., D. Schneider, S. Carter, and T. White. 1987. "Paradoxical Effects of Thought Suppression." *Journal of Personality and Social Psychology* 53: 5–13.

Welter, M. L., P. Burbaud, S. Fernandez-Vidal, et al. 2011. "Basal Ganglia Dysfunction in OCD: Subthalamic Neuronal Activity Correlates with Symptom

Severity and Predicts High-Frequency Stimulation Efficacy." *Translational Psychiatry* 1: e5.

Whittal, M. L., D. S. Thodarson, and P. D. McLean. 2005. "Treatment of Obsessive-Compulsive Disorder: Cognitive Behavior Therapy vs. Exposure and Response Prevention." *Behaviour Research and Therapy* 43: 1559–1576.

Wilson, R. 2016. *Stopping the Noise in Your Head: The New Way to Overcome Anxiety & Worry.* Deerfield Beach, FL: Health Communications, Inc.

Yang, C. C., A. Barros-Loscertales, D. Pinazo, et al. 2016. "State and Training Effects of Mindfulness Meditation on Brain Networks Reflect Neuronal Mechanisms of Its Antidepressant Effect." *Neural Plasticity*: Article ID 9504642.

Yoo, S., N. Gujar, P. Hu, F. A. Jolesz, and M. P. Walker. 2007. "The Human Emotional Brain Without Sleep—A Prefrontal Amygdala Disconnect." *Current Biology* 17: 877–878.

Zeidan, F., K. T. Martucci, R. A. Kraft, J. G. McHaffie, and R. C. Coghill. 2013. "Neural Correlates of Mindfulness Meditation-Related Anxiety Relief." *Social Cognitive and Affective Neuroscience* 9: 751–759.

Zelano, C., H. Jiang, G. Zhou, et al. 2016. "Nasal Respiration Entrains Human Limbic Oscillations and Modulates Cognitive Function." *The Journal of Neuroscience* 36: 12448–12467.

Ziemann, A. E., J. E. Allen, N. S. Dahdaleh, et al. 2009. "The Amygdala Is a Chemosensor That Detects Carbon Dioxide and Acidosis to Elicit Fear Behavior." *Cell* 139: 1012–1021.

Catherine M. Pittman, PhD, is a licensed clinical psychologist specializing in the treatment of anxiety disorders and brain injuries. She is professor of psychology at Saint Mary's College in Notre Dame, IN; where she has taught for over thirty years.

William H. Youngs, PhD, is a licensed clinical psychologist with a private practice in clinical neuropsychology in the greater South Bend, IN, area. He served as a clinical neuropsychologist with Memorial Hospital of South Bend for twenty-five years, and a visiting assistant professor of psychology at Saint Mary's College in Notre Dame, IN; where he has taught undergraduate courses in neuropsychology, cognitive psychology, psychological assessment, abnormal psychology, and theories of personality.

MORE BOOKS from
NEW HARBINGER PUBLICATIONS

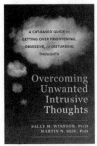
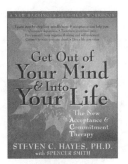
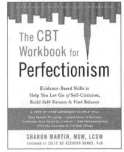

FROM OUR PUBLISHER—

As the publisher at New Harbinger and a clinical psychologist since 1978, I know that emotional problems are best helped with evidence-based therapies. These are the treatments derived from scientific research (randomized controlled trials) that show what works. Whether these treatments are delivered by trained clinicians or found in a self-help book, they are designed to provide you with proven strategies to overcome your problem.

Therapies that aren't evidence-based—whether offered by clinicians or in books—are much less likely to help. In fact, therapies that aren't guided by science may not help you at all. That's why this New Harbinger book is based on scientific evidence that the treatment can relieve emotional pain.

This is important: if this book isn't enough, and you need the help of a skilled therapist, use the following resources to find a clinician trained in the evidence-based protocols appropriate for your problem. And if you need more support—a community that understands what you're going through and can show you ways to cope—resources for that are provided below, as well.

Real help is available for the problems you have been struggling with. The skills you can learn from evidence-based therapies will change your life.

Matthew McKay, PhD
Publisher, New Harbinger Publications

**If you need a therapist, the following organization
can help you find a therapist trained in cognitive behavioral therapy (CBT).**

The Association for Behavioral & Cognitive Therapies (ABCT) Find-a-Therapist service offers a list of therapists schooled in CBT techniques. Therapists listed are licensed professionals who have met the membership requirements of ABCT and who have chosen to appear in the directory.

Please visit www.abct.org and click on *Find a Therapist*.

**For additional support for patients, family, and friends,
please contact the following:**

International OCD Foundation (IOCDF)
Visit www.ocfoundation.org

Anxiety and Depression Association of American (ADAA)
please visit www.adaa.org

Register your **new harbinger** titles for additional benefits!

When you register your **new harbinger** title—purchased in any format, from any source—you get access to benefits like the following:

- Downloadable accessories like printable worksheets and extra content

- Instructional videos and audio files

- Information about updates, corrections, and new editions

Not every title has accessories, but we're adding new material all the time.

Access free accessories in 3 easy steps:

1. Sign in at NewHarbinger.com (or **register** to create an account).

2. Click on **register a book**. Search for your title and click the **register** button when it appears.

3. Click on the **book cover or title** to go to its details page. Click on **accessories** to view and access files.

That's all there is to it!

If you need help, visit:

NewHarbinger.com/accessories

new harbinger
CELEBRATING
40 YEARS